Understanding Search Engines

SOFTWARE · ENVIRONMENTS · TOOLS

The series includes handbooks and software guides, as well as monographs on practical implementation of computational methods, environments, and tools. The focus is on making recent developments available in a practical format to researchers and other users of these methods and tools.

Editor-in-Chief

Jack J. Dongarra
University of Tennessee and Oak Ridge National Laboratory

Editorial Board

James W. Demmel, *University of California, Berkeley*
Dennis Gannon, *Indiana University*
Eric Grosse, *AT&T Bell Laboratories*
Ken Kennedy, *Rice University*
Jorge J. Moré, *Argonne National Laboratory*

Software, Environments, and Tools

Lloyd N. Trefethen, *Spectral Methods in MATLAB*

E. Anderson, Z. Bai, C. Bischof, S. Blackford, J. Demmel, J. Dongarra, J. Du Croz, A. Greenbaum, S. Hammarling, A. McKenney, and D. Sorensen, *LAPACK Users' Guide, Third Edition*

Michael W. Berry and Murray Browne, *Understanding Search Engines: Mathematical Modeling and Text Retrieval*

Jack J. Dongarra, Iain S. Duff, Danny C. Sorensen, and Henk A. van der Vorst, *Numerical Linear Algebra for High-Performance Computers*

R. B. Lehoucq, D. C. Sorensen, and C. Yang, *ARPACK Users' Guide: Solution of Large-Scale Eigenvalue Problems with Implicitly Restarted Arnoldi Methods*

Randolph E. Bank, *PLTMG: A Software Package for Solving Elliptic Partial Differential Equations, Users' Guide 8.0*

L. S. Blackford, J. Choi, A. Cleary, E. D'Azevedo, J. Demmel, I. Dhillon, J. Dongarra, S. Hammarling, G. Henry, A. Petitet, K. Stanley, D. Walker, and R. C. Whaley, *ScaLAPACK Users' Guide*

Greg Astfalk, editor, *Applications on Advanced Architecture Computers*

Françoise Chaitin-Chatelin and Valérie Frayssé, *Lectures on Finite Precision Computations*

Roger W. Hockney, *The Science of Computer Benchmarking*

Richard Barrett, Michael Berry, Tony F. Chan, James Demmel, June Donato, Jack Dongarra, Victor Eijkhout, Roldan Pozo, Charles Romine, and Henk van der Vorst, *Templates for the Solution of Linear Systems: Building Blocks for Iterative Methods*

E. Anderson, Z. Bai, C. Bischof, J. Demmel, J. Dongarra, J. Du Croz, A. Greenbaum, S. Hammarling, A. McKenney, S. Ostrouchov, and D. Sorensen, *LAPACK Users' Guide, Second Edition*

Jack J. Dongarra, Iain S. Duff, Danny C. Sorensen, and Henk van der Vorst, *Solving Linear Systems on Vector and Shared Memory Computers*

J. J. Dongarra, J. R. Bunch, C. B. Moler, and G. W. Stewart, *Linpack Users' Guide*

Understanding Search Engines
Mathematical Modeling
and Text Retrieval

Michael W. Berry

University of Tennessee
Knoxville, Tennessee

Murray Browne

University of Tennessee
Knoxville, Tennessee

SOFTWARE · ENVIRONMENTS · TOOLS

Society for Industrial and Applied Mathematics
Philadelphia

Library of Congress Cataloging-in-Publication Data

Berry, Michael W.
    Understanding search engines : mathematical modeling and text retrieval / Michael W. Berry, Murray Browne.
        p. cm. -- (Software, environments, tools)
    Includes bibliographical references.
    ISBN 0-89871-437-0 (pbk.)
    1. Web search engines. 2. Vector spaces 3. Text processing (Computer science)
I. Browne, Murray. II. Title. III. Series.
TK5105.884.B47 1999
025.04 -- dc21   99-30649

ISBN 0-89871-437-0

siam is a registered trademark.

To our wives (Teresa and Shirley) and daughters (Amanda, Rebecca, Cynthia, and Bonnie).

Contents

Preface

Anyone who has used a Web search engine with any regularity knows that there is an element of the unknown with every query. Sometimes the user will type in a *stream-of-consciousness* query and the documents retrieved are a perfect match, while the next query can be seemingly succinct and focused only to earn the bane of all search results — the *no documents found* response. Oftentimes the same queries can be submitted on different databases with just the opposite results. It is an experience aggravating enough to make one swear off doing Web searches as well as swear at the developers of such systems.

However, because of the transparent nature of computer software design, there is a tendency to forget the decisions and trade-offs that are constantly made throughout the design process affecting the performance of the system. One of the main objectives of this book is to identify to the novice search engine builder — such as the senior level computer science or applied mathematics student or the information sciences graduate student specializing in retrieval systems — the impact of certain decisions that are made at various junctures of this development. One of the major decisions in developing information retrieval systems is selecting and implementing the computational approaches within an integrated software environment. Applied mathematics plays a major role in search engine performance and *Understanding Search Engines* (or USE) focuses on this area, bridging the gap between the fields of applied mathematics and information management, disciplines that previously have operated largely in independent domains.

But this book does not only fill the gap between applied mathematics and information management, it also fills a niche in the information retrieval literature. The work of William Frakes and Ricardo Baeza-Yates (eds.), *Information Retrieval: Data Structures & Algorithms*, a 1992 collection of journal articles on various related topics, and Gerald Kowalski's (1997) *Information Retrieval Systems: Theory and Implementation*, a broad overview of information retrieval systems, are fine textbooks on the topic, but both understandably lack the gritty details of the mathematical computations needed to build more successful search engines.

With this in mind, this book does not provide an overview of information retrieval systems, but prefers to assume a supplementary role to the above-mentioned books. Many of the ideas we use here were presented and developed as part of a Data and Information Management course at the University of Tennessee's Computer Science Department, a course which won the 1997 Undergraduate Computational Engineering and Science Award sponsored by the United States Department of Energy and the Krell Institute. The course, which required student teams to build their own search engines, has provided invaluable background material in the development of this book.

As mentioned earlier, this book concentrates on the applied mathematics portion of search engines. Although not transparent to the pedestrian search engine user, mathematics plays an integral part in information retrieval systems by computing the emphasis the query terms have in their relationship to the database. This is especially true in vector space modeling, which is one of the predominant techniques used in search engine design. With vector space modeling, traditional orthogonal matrix decompositions from linear algebra can be used to encode both term and documents in k-dimensional space.

However, that is not to say that other computational methods are not useful or valid, but in order to teach future developers the intricate details of a system, a single approach had to be selected. Therefore, the reader can expect a fair amount of math, including explanations of algorithms and data structures and how they operate in information retrieval systems. This book will not hide the math (concentrated in chapters 3 and 4), nor will it allow itself to get bogged down in it either. A person with a nonmathematical background (such as an information scientist) can still appreciate some of the mathematical intricacies involved with building search engines without reading the more technical chapters 3 and 4.

To maintain its focus on the mathematical approach this book has purposely avoided digressions into Java programming, HTML programming, and how to create a web interface. An informal conversational approach has been adopted to give the book a less intimidating tone, which is especially important considering the possible multidisciplinary backgrounds of its potential readers; however, standard math notation will be used. Boxed items throughout the book contain ancillary information, such as mathematical examples, anecdotes, and current practices, to help guide the discussion. Websites providing software (e.g., CGI scripts, text parsers, numerical software) and text corpora are provided in chapter 9.

Acknowledgments

The authors would like to gratefully acknowledge the support and encouragement of SIAM, the United States Department of Energy, the Krell Institute, the National Science Foundation for supporting related research, the University of Tennessee, the students of CS460/594 (fall semester 1997), and graduate assistant Luojian Chen. Special thanks to Alan Wallace and David Penniman from the School of Information Sciences at the University of Tennessee, Padma Raghavan and Ethel Wittenberg in the Department of Computer Science at the University of Tennessee, Barbara Chen at H.W. Wilson Company, and Martha Ferrer at Elsevier Science SPD for their helpful proofreading, comments, and/or suggestions. The authors would also like to thank Katie Terpstra and Eric Clarkson for their work with the book cover artwork and design, respectively.

Hopefully, this book will help future developers, whether they be students or software engineers, to lessen the aggravation encountered with the current state of search engines. It is a critical time for search engines and the future of the Web itself, as both ultimately depend on how easily users can find the information they are looking for.

MICHAEL W. BERRY
MURRAY BROWNE

DOONESBURY ©G.B. Trudeau. Reprinted with permission of UNIVERSAL PRESS
SYNDICATE. All rights reserved.

Chapter 1

Introduction

We expect a lot from our search engines. We ask them vague questions
about topics that we're unfamiliar with ourselves and in turn anticipate a
concise, organized response. We type in *principal* when we meant *principle*.
We incorrectly type the name *Lanzcos* and fully expect the search engine to
know that we really meant *Lanczos*. Basically we are asking the computer
to supply the information we want, instead of the information we asked for.
In short, users are asking the computer to reason intuitively. It's a tall order
and in some search systems you would probably have better success if you
laid your head on the keyboard and coaxed the computer to try to read your
mind.

Of course these problems are nothing new to the reference librarian who
works the desk at a college or public library. An experienced reference librarian knows that a few moments spent with a patron, listening, asking
questions, and listening some more can go a long way in efficiently direct-
ing the user to the source that will fulfill the user's information needs. In
the computerized world of searchable databases this same strategy is being
developed, but it has a long way to go before being perfected.

There is another problem with locating the relevant documents for a
respective query, and that is the increasing size of collections. Heretofore,
the focus of new technology has been more on processing and digitizing
information, whether it be text, images, video, or audio, than on organizing
it. It has created a situation information designer Richard Saul Wurman

[64] refers to as a *tsunami of data*:

> This is a tidal wave of unrelated, growing data formed in bits
> and bytes, coming in an unorganized, uncontrolled, incoherent
> cacophony of foam. It's filled with flotsam and jetsam. It's filled
> with the sticks and bones and shells of inanimate and animate
> life. None of it is easily related, none of it comes with any orga-
> nizational methodology.

To combat this tsunami of data, search engine designers have developed
a set of mathematically based tools that will improve search engine performance. Such tools are invaluable for improving the way in which terms
and documents are automatically synthesized. Term weighting methods, for
example, are used to place different emphases on a term's (or a key word's)
relationship to the other terms and other documents in the collection. One
of the most effective mathematical tools embraced in automated indexing is
the vector space model [51].

In the vector space information retrieval (IR) model, a unique vector is
defined for every term in every document. Another unique vector is computed for the user's query. With the queries being easily represented in the
vector space model, searching translates to the computation of distances between query and document vectors. However, before vectors can be created
in the document, some preliminary document preparation must be done.

1.1   Document File Preparation

Librarians are well aware of the necessities of organizing and extracting
information. Through decades (really, centuries) of experience, librarians
have refined a system of organizing materials that come into the library.
Every item is catalogued based on some individual's or group's assessment
of what that book is about, followed by appropriate entries in the library's
on-line or card catalog. Although it is often outsourced, essentially each
book in the library has been individually indexed or reviewed to determine
its contents. This approach is generally referred to as *manual indexing*.

1.1.1   Manual Indexing

As with most approaches, there are are some real advantages and disadvantages to manual indexing. One major advantage is that a human indexer can

establish relationships and concepts between seemingly different topics that can be very useful to future readers. Unfortunately, this task is expensive and time consuming, and can be subject to the background and personality of the indexer. For example, studies by Cleverdon [17] reported that if two groups of people construct thesauri in a particular subject area, the overlap of index terms was only 60%. Furthermore, if two indexers used the same thesaurus on the same document, common index terms were shared in only 30% of the cases.

Also of potential concern is that the manually indexed system may not be reproducible or if the original system was destroyed or modified it would be difficult to re-create. All in all, it is a system that has worked very well, but with the proliferation of digitized information on the World Wide Web (WWW), there is a need for a more *automated* system.

Fortunately because of increased computer processing power in this decade, computers have been used to extract and index words from documents in a more automated fashion. This has also changed the role of manual subject indexing. According to Kowalski [33, p. 50], "The primary use of manual subject indexing now shifts to the abstraction of concepts and judgments on the value of the information."

Of course, the next stage in the evolution of automatic indexing is being able to link related concepts even when the query doesn't specifically make such a request.

1.1.2   File Cleanup

One of the least glamorous and often overlooked parts of search engine design is the preparation of the documents that are going to be searched. A simple analogy might be the personal filing system you may have in place at home. Everything from receipts to birth certificates to baby pictures are thrown into a filing cabinet or a series of boxes. It's all there, but without file folders, plastic tabs, color coding, or alphabetizing it's nothing more than a heap of paper. Subsequently, when you go to search for the credit card bill you thought you paid last month, it's an exercise similar to rummaging through a wastebasket.

There is little difference between the previously described home filing system and documents in a web-based collection, especially if nothing has been done to *standardize* those documents to make them searchable. In other words, unless documents are cleaned up or purified by performing pedestrian tasks such as making sure every document has a title, marking where each document begins and ends, and handling parts of the documents that are not text (such as images), then most search engines will respond by returning the wrong document(s) or fragments of documents.

One misconception is that information that has been formatted through an HTML (hypertext markup language) editor and displayed in a browser is sufficiently formatted, but that is not always the case because HTML was designed as a platform-independent language. In general, Web browsers are very forgiving with built-in error recovery and thus will display almost any kind of text, whether it looks good or not. However, search engines have more stringent format requirements and that is why when building a web-based document collection for a search engine, each HTML document has to be validated into a more specific format prior to any indexing.

1.2   Information Extraction

In chapter 2, we will go into more detail on how to go about doing this cleanup, which is just the first of many procedures needed for what is referred to as *item normalization*. We also look at how the words of a document are *processed* into searchable tokens by addressing such areas as processing tokens and stemming. Once these prerequisites are met, the documents are ready to be indexed.

1.3   Vector Space Modeling

SMART (System for the Mechanical Analysis and Retrieval of Text), developed by Gerald Salton and his colleagues at Cornell University [51], was one of the first examples of a vector space IR model. In such a model, both terms and/or documents are encoded as vectors in k-dimensional space. The choice k can be based on the number of unique terms, concepts, or perhaps classes associated with the text collection. Hence, each vector component (or dimension) is used to reflect the importance of the corresponding term/concept/class in representing the semantics or meaning of a document.

Figure 1.1 demonstrates how a simple vector space model can be represented as a *term-by-document* matrix. Here, each column defines a document while each row corresponds to a unique term or keyword in the collection.

The values stored in each matrix element or cell define the frequency that a term occurs in a document. For example, *Term* 1 appears once in both *Document* 1 and *Document* 3 but not in the other two documents (see Figure 1.1). Figure 1.2 demonstrates how each column of the 3×4 matrix in Figure 1.1 can be represented as a vector in 3-dimensional space. Using a k-dimensional space to represent documents for clustering and query matching purposes can become problematic if k is chosen to be the number of terms (rows of matrix in Figure 1.1). Chapter 3 will discuss methods for representing term-document associations in lower dimensional vector spaces and how to construct term-by-documents using term-weighting methods [20, 49, 58] to show the importance a term can have within a document or across the entire collection.

	Document 1	Document 2	Document 3	Document 4
Term 1	1	0	1	0
Term 2	0	0	1	1
Term 3	0	1	1	0

Figure 1.1: Small term-by-document matrix.

Through the representation of queries as vectors in the k-dimensional space, documents (and terms) can be compared and ranked according to similarity with the query. Measures such as the Euclidean distance and cosine of the angle made between document and query vectors provide the similarity values for ranking. Approaches based on conditional probabilities (logistic regression, Bayesian models) to judge document-to-query similarities are not the scope of this book; however, references to other sources such as [24, 25] have been included.

1.4 Matrix Decompositions

In simplest terms, search engines take the user's query and find all the documents that are related to the query. However, this task becomes complicated quickly, especially when the user wants more than just a literal match. One approach known as *latent semantic indexing* or LSI [7, 18] attempts to do more than just literal matching. Employing a vector space representation of both terms and documents, LSI can be used to find relevant documents that may not even share any search terms provided by the user. Modeling

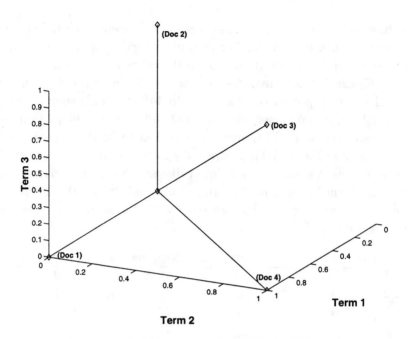

Figure 1.2: Representation of documents in a 3-dimensional vector space.

the underlying term-to-document association patterns or relationships is the key for *conceptual*-based indexing approaches such as LSI.

The first step in modeling the relationships between the query and a document collection is just to keep track of which document contains which terms or which terms are found in which documents. This is a major task requiring computer-generated data structures (such as term-by-document matrices) to keep track of these relationships. Imagine a spreadsheet with every document of a database arranged in columns. Down the side of the chart is a list of all the possible terms (or words) that could be found in those documents. Inside the chart, rows of integers (or perhaps just ones and zeros) mark how many times the term appears in the document (or if it appears at all).

One interesting characteristic of term-by-document matrices is that they usually contain a greater proportion of zeros; i.e., they are quite *sparse*. Since every document will contain only a small subset of words from the dictionary, this phenomenon is not too difficult to explain. On the average, only about 1% of all the possible elements or cells are populated [7, 9, 31].

When a user enters a query, the retrieval system (search engine) will attempt to extract all matching documents. Recent advances in hardware technologies have produced extremely fast computers, but these machines are not so fast that they can scan through an entire database every time the user makes a query. Fortunately, through the use of concepts from applied mathematics, statistics, and computer science, the actual amount of information that must be processed to retrieve useful information is continuing to decrease. But such reductions are not always easy to achieve, especially if one wants to obtain more than just a literal match.

Efficiency in indexing via vector space modeling requires special encodings for terms and documents in a text collection. The encoding of term-by-document matrices for lower dimensional vector spaces (where the dimension of the space is much smaller than the number of terms or documents) using either continuous or discrete matrix decompositions is required for LSI-based indexing. The singular value decomposition (SVD) [26] and semidiscrete decomposition (SDD) [31] are just two examples of the various matrix decompositions arising from numerical linear algebra that can be used in vector space IR models such as LSI. The matrix factors produced by these decompositions provide automatic ways of encoding or representing terms and documents as vectors in any dimension. The clustering of similar or related terms and documents is realized through probes into the derived vector space model, i.e., queries. A more detailed discussion of the use of matrix decompositions such as the SVD and SDD for IR models will be provided in chapter 4.

1.5 Query Representations

Query matching within a vector space IR model can be very different from conventional item matching. Whereas with the latter we think of a user typing in a few terms and the search engine matching the user's terms to those indexed from the documents in the collection, in vector space models such as LSI, the query can be interpreted as another (or new) document. Upon submitting the query, the search engine will retrieve the cluster of documents (and terms whose word usage patterns reflect that of the query).

This difference is not necessarily transparent to the experienced searcher. Those trained in searching are often taught Boolean searching methods (especially in library and information sciences), i.e., the connection of search terms by *and* and *or*. For example, if a Boolean searcher queries a CD-ROM encyclopedia on *German shepherds and bloodhounds*, the documents retrieved must have information about both German shepherds and bloodhounds. In a pure Boolean search, if the query was *German shepherds or bloodhounds*, the documents retrieved will include any article that has something about German shepherds or bloodhounds.

IR models can differ in how individual search terms are processed. Typically, all terms are treated equally with insignificant words removed. However, some terms may be *weighted* according to their importance. Oddly enough, with vector space models, the user may be better off listing as many relevant terms as he or she can in the query, in contrast to a Boolean user who usually types in just a few words. In vector space models, the more terms that are listed, the better chance the search engine has in finding similar documents in the database.

Natural language queries such as "I would like articles about German shepherds and bloodhounds," compose yet another form of query representation. Even though to the trained Boolean searcher this seems unnatural, this type of query can be easier and more accurate to process, because the importance of each word can be gauged from the semantic structure of the sentence. By discarding insignificant words (such as *I, would, like*) a conceptual-based IR system is able to determine which words are more important and therefore should be used to extract clusters of related documents and/or terms.

In chapter 5, we will further discuss the process of query binding or how the search engine takes abstract formulations of queries and forms specific requests.

1.6   Ranking and Relevance Feedback

As cold as it sounds, search engine users really don't care how a search engine works; they are just interested in getting the information they've requested. Once they have the answer they want, they log off — end of query. This disregarding attitude creates certain challenges for the search engine builder. For example, only the user can ultimately judge if the retrieved information meets his or her needs. In information retrieval, this is known as *relevance*, or judging how well the information received matches the query. (Augmenting this problem is the fact that oftentimes the user is not sure what he or she is looking for.) Fortunately, vector space modeling, because of its applied mathematical underpinnings, has characteristics that improve the chances that the user will eventually receive relevant documents to his or her corresponding query. The search engine does this in two ways: by ranking the retrieved documents according to how well they match the query and relevance feedback or by asking the user to identify which documents best meet his or her information needs, and then based on that answer, resubmitting the query.

Applied mathematics plays such an integral part of vector-based search engines because there is already in place a quantifiable way to say, *Document A ranks higher in meeting your criteria based on your search terms than Document B.* This idea can then be taken one step further, when the user is asked, *Do you want more documents like Document A or Document B or Document C... ?* After the user makes the selection, more similar documents are retrieved and ranked. Again, this process is known as relevance feedback.

Using a more mathematical perspective, we will discuss in chapter 6 the use of vector-based similarity measures (e.g., cosines) to rank-order documents (and terms) according to their similarity with a query.

1.7   User Interface

If users really don't care much about how search engines actually work, what really does matter to them? Only the search results? This is probably true, but not necessarily, because even the best search engine ever built, guaranteed to produce great amounts of relevant documents for every query, may be very difficult to access. Hence, users have a tendency to avoid them. Conversely, how often have we returned to some ineffectual search engine, just because it is easy to use? Probably more times than we are willing to admit.

These two extreme examples illustrate the importance of the user interface in search engine design. On the Web, a user typically fills out a simple, short form and then submits his or her query. But does the user know whether to type in a few words, Boolean operators, different types of spellings of the same words, or whether the query should be in the form of a question? What kind of help instructions should be included and displayed?

Another factor related to the user interface is how the retrieved documents will be displayed. Will it be titles only, titles and abstracts, or entire documents with the query terms highlighted (known in information retrieval lexicon as KWIC or *keyword in context*). Then there is the issue of speed. Should the user ever be expected to wait for results more than 10 seconds after pressing the return key? In the design of search engines, there are trade-offs that will affect the speed of the retrieval. Those trade-offs will be discussed in more depth in chapter 7 included with other general dos and don'ts of interface design.

1.8   Course Project

As part of a recent course entitled *Data and Information Management* at the University of Tennessee (fall 1997), students applied the modular approach described in this book to design their own search engines. In chapter 8, we describe how this project was organized and managed for a modest number of students (approximately 25). Sharing some of the real classroom experiences students had in the design, testing, and presentation phase of this project should help instructors and students construct similar projects for courses in computer science and information science.

1.9   Final Comments

Before we begin going into depth about each of the interrelated ingredients that goes into building a search engine, we want to remind the reader why the book is formatted the way it is. We are anticipating the likelihood that interested readers will have different backgrounds and viewpoints about search engines. Therefore, we purposely tried to separate the nontechnical material from the mathematical calculations. Those with an information sciences or nonmathematical background should consider skimming or skipping chapters 3 and 4 and sections 6.1 and 6.2. However, we encourage those with applied mathematics or computer science backgrounds to read the less-technical chapters 1, 5, 6, 7, and 8, because the exposure to the information science *perspective* of search engines is critical for both assessing performance and understanding how users *see* search engines. In chapter 9, we list background sources (including current websites) that not only have been influential in writing this book, but can also provide opportunities for further understanding.

Another point worth reminding readers about is that vector space modeling was chosen for conceptual IR to demonstrate the important role that applied mathematics can play in communicating new ideas and attracting multidisciplinary research in the design of intelligent search engines. Certainly there are other IR approaches, but we hope our experiences with the vector space model will help pave the way for future system designers to build better, more useful search engines.

Chapter 2

Document File Preparation

As mentioned briefly in the introduction, a major part of search engine development is making decisions about how to prepare the documents that are to be searched. If the documents are automatically indexed, they will be managed/handled much differently than if they were just manually indexed. The search engine designer must be aware that building the automatic index is as important as any other component of the search engine development.

As pointed out by Korfhage in [32], system designers must also take into consideration that it is not unusual for a user to be linked into many different databases through a single user interface. Each one of these databases often will have its own way of handling the data. Also, the user is unaware that he or she is searching different databases (nor do they care). It's up to the search engine builder to smooth over these differences and make them transparent to the user.

Building an index comprises two lengthy steps:

1. Document analysis and, for the lack of a better word, *purification*. This requires analyzing how each document in each database (e.g., Web documents) is organized in terms of what makes up a document (title, author or source, body) and how the information is presented. Is the critical information in the text or is it presented in tables, charts, graphics, or images? Decisions must be made on what information or parts of the document (referred to as *zoning* [33]) will be indexed and which information will not.

2. Token analysis or term extraction. On a word-by-word basis, decision must be made on which words (or phrases) should be used as

referents in order to best represent the semantic content (meaning) of documents.

2.1   Document Purification and Analysis

After even limited exposure to the WWW, one quickly realizes that HTML documents are composed of many attributes such as graphic images, photos, tables, charts, and audio clips, and those are just the visible characteristics. By viewing the source code of an HTML document, one also sees a variety of tags such as <TITLE>, <COMMENT>, and <META> tags, which also describe features that help identify how the document is organized and displayed. Obviously, hypertext documents are *more* than just text.

Any search engine on the WWW must address the heterogeneity of HTML documents. By examining the limitations of the various commercial search engines, one can see the nonuniformity of processing HTML documents. It is not uncommon to simply avoid indexing nontextual information altogether.

According to [42] and [59], the following features are often ignored in many current search engines on the Web:

- <COMMENT> tags, which allow the page developer to leave instructions or reminders about the page hidden from view.

- the <ALT TEXT> attribute, which allows the page developer to provide a text description of an image in case the user has the browser set to *text only.*

- <META> tags, which (in most cases) are chosen by the page developer to describe the page. Until recently, some search engines (e.g., Excite) deliberately avoided indexing <META> tags to avoid problems/biases associated with the ranked returned list of documents for specific queries.

- image maps, frames, and in some cases universal resource locators (URLs) are not indexed. URLs are usually defined within <HREF> tags.

2.1.1   Text Formatting

Before moving from step 1 of analyzing a document to step 2 of processing its individual elements, it is critical that each document be in ASCII or some similar (editable) format. This seems like a *standard* requirement, but one must remember that some documents are added into collections by optical character reader (OCR) scanners and may be recast in formats such as postscript. Such a format restricts searching because it exists more as an *image* rather than a collection of individual, searchable elements. Documents can be converted from postscript to ASCII files, even to the point that special and critical elements of the document such as the title, author, and date can be flagged and processed appropriately [32].

Search engine developers must also determine how they are going to index the text. Later in the next section, we will discuss the process of item normalization (use of stop lists, stemmers, and the like), which is typically performed after the search engine has selected which text to index. Again, judging by the way vendors describe their own search engines [42, 59], not all text is equal either.

2.1.2 Validation

Producing *valid* HTML files, unfortunately, is not as straightforward as one would expect. The lack of consistent tagging (by users) and nonstandard software for HTML generation produces erroneous webpages that can make both parsing and displaying a nightmare. On-line validation services, however, are making a difference in that users can submit their webpage(s) for review and perhaps improve their skills in HTML development. An excellent resource for webpage/HTML validation is provided by the W3C HTML Validation Service at `http://validator.w3.org/`. Users can submit the URL of their webpage for validation. Some standards for HTML syntax are now being implemented by various computer vendors (e.g., IBM, Microsoft, Netscape, Novell, Spyglass, and Sun Microsystems) as part of the World Wide Web Consortium or W3C (see `http://www.w3.org`). To identify (or specify) which version of HTML is used within a particular webpage, the formal public identifier (FPI) comment line is commonly used. A sample FPI declaration for the W3C 3.2 HTML syntax is provided in Figure 2.1.

In general, search engines on the Web only extract text (excluding punctuation) from the title tags, the header tags (`<H1>` and `<H2>`) and the first characters of the file, which can range from the first 275 characters to the first 70,000 characters. Also, the search engine developer may select the first 100 significant words or the first 20 lines/records (which could contain a code/script rather than text). Remember that search engines basically

```
<!-- The first non-comment line is the FPI declaration for -->
<!-- this document. It declares to a parser and a human     -->
<!-- reader the HTML syntax used in its creation. In this   -->
<!-- example, the Document Type Definition (DTD) from the   -->
<!-- W3C for HTML 3.2 was used during creation. It also     -->
<!-- dictates to a parser what set of syntax rules to use   -->
<!-- when validating.                                        -->
<!DOCTYPE HTML PUBLIC "-//W3C//DTD HTML 3.2//EN">
<HTML>
  <HEAD>
    <TITLE>Example</TITLE>
  </HEAD>
<BODY>
 ...
</BODY>
</HTML>
```

Figure 2.1: FPI for HTML conforming to the W3C 3.2 standard.

work from a *representation* of the WWW, not the entire thing.

To counter the *spamming* technique[1] of making the text color the same as the background, some search engines purposely ignore this *invisible* text. In some cases, text of a smaller font is ignored as well as any words that contain numbers.

2.2   Manual Indexing

Manual indexing, or indexing that is done by a person, connotes visions of dedicated men and women willingly locked into small windowless rooms, sitting at sparse desks hunched over stacks of papers, reading and underlining key words for the index. On coffee breaks, the indexers gather in the small break room and debate the nuances of language and exchange anecdotes about the surprising relationships between seemingly incongruent subjects.

Considering the proliferation of digital sources such as the WWW, with its 320 million and growing indexable pages [36], you'd think that manual

---

[1]Technique used by webpage designers to improve the overall ranking of their page compared with others for targeted search strings (queries).

indexers must have gone the way of bank tellers and full service gas station attendants. On the contrary, there are still major players in the information industry who, despite the labor costs of having documents analyzed individually, coupled with the inherent problems of consistency throughout a group of catalogers, still prefer using a degree of manual indexing. Some examples include:

- *Yahoo!* Even though Yahoo! is technically more of a subject guide than a search engine, the collection is manually indexed. Instead of a web crawler looking at sites and then pulling back information for indexing, web masters submit URLs for Yahoo! to peruse. If the Yahoo! catalogers like the site and it fits into the Yahoo! directory in terms of content (only about 30% of requests to Yahoo! are accepted), then the site is indexed and someone decides which category to place the website. However, according to [59], it's not entirely without shortcomings, "... as sites cannot be adequately summarized by being listed in only one or two categories."

  According to [60], another characteristic of subject guide type databases in general is that they have smaller databases than more automatic search engines, and have a tendency to index the top level of a website. On the positive side, "... because their maintenance includes human intervention, subject guides greatly reduce the probability of retrieving results out of context."

  In other words, disadvantages of higher costs and less coverage are balanced by the manual indexer's ability to recognize relationships in seemingly unrelated documents.

- *EMBASE.* Although Elsevier Science's bibliographic database (EMBASE) covering pharmacology and biomedicine experimented with the more cost-effective and efficient methods of machine-aided indexing, their goal was to have those methods work in tandem with manual indexers. According to Martha Ferrer, User Services Manager at Elsevier Science SPD, Elsevier's "aim is not to replace the *human* indexer, [but] rather to assist him/her in the indexing process. To that aim, the use of machine-aided indexing is fast becoming standard practice."

- *National Library of Medicine.* The National Library of Medicine, publishers of MeSH, the Medical Subject Headings [46], uses indexers to "assign as many headings as necessary (usually 10–12) to characterize accurately the content of a journal article."

- *H. W. Wilson Company.* H. W. Wilson Company, publishers of the *Reader's Guide to Periodical Literature* and other indexes, places major importance on having a person assigning subject headings. Manual indexers assign the key words in all their index publications. According to H. W. Wilson's Director of Indexing Barbara Chen, "No search engine machine-aided indexing can do what a human can do... being able to understand related concepts are important."

For the information industry to make such an effort and expense (in contrast to the emerging focus on automatic indexing) to manually assign terms indicates the importance information professionals place on being able to recognize relationships and concepts in documents. It's this ability to identify the broader, narrower, and related subjects that keeps manual indexing a viable alternative in an industry that is somewhat dominated by automatic indexing. It also provides a goal for the automatic indexing system of being able to accurately forge relationships between documents that on the surface are not lexically linked.

2.3 Automatic Indexing

Automatic indexing or using algorithms/software to extract terms for indexing is the predominant method for processing documents from large web databases. In contrast to the connotation of manual indexers being holed up in their windowless rooms, the vision of automatic indexes consists of huge automatic computerized robots crawling throughout the Web all day and night, collecting documents and indexing every word in the text. This latter vision is probably as overly romanticized as the one for the manual indexers, especially considering that the robots are stationary and make requests to the servers in a fashion similar to a user making a request to a server.

Another difference between manual and automatic indexing is that concepts are realized in the indexing stage as opposed to document extraction/collection. This characteristic of automatic indexing places additional pressure on the search engine builder to provide some means of searching for broader, narrower, or related subjects. Ultimately, though, the goal of each system is the same: to extract from the documents the words that will allow a searcher to find the best documents to meet his or her information needs.

Industrial Search Engines

By looking at the features of some of major search engines such as AltaVista, Infoseek, and Excite we can get a general idea of how major search engines do their automatic indexing. It can also offer insights on the types of decisions a search engine builder must make when extracting index terms [59].

Each search engine usually has its own *crawler*, which is constantly indexing the Web. Such crawlers may index between 3 to 10 million pages per day. While some crawlers just randomly search, others are specifically looking at previously indexed pages for updated information or are guided by user submissions. One popular measurement of search engine viability is tracking how many days it takes for a webmaster to make an *index submission* before the crawler actually visits (some search engines claim a one-day turnaround, while others say it could be as long as a month).

While studying automatic indexing, keep in mind that the search engine or crawler *grabs* only part of the webpage and copies it into a more localized database. This means when a user submits a query to the search engine, only that particular search engine's representation (or subset) of the Web is actually searched. Only then are you directed to the appropriate (current) URL. This explains in part why links from search results are invalid or redirected and explains the importance of search engines to update and refresh their existing databases.

However, there are limits to what search engines are willing or able to index. For example, the Excite web crawler grabs everything it can pull back in 30 seconds, whereas Lycos takes the first 275 characters on the page, and Infoseek will take only the first 16 kilobytes of a webpage and use it for the index [59].

Industrial Search Engines, contd.

Once webpages are *pulled in*, guidelines are set in advance on what exactly to index. Since most of the documents are written in HTML, each search engine must decide what to do with frames, password protected sites, comments, metatags, and images. Search engine designers must determine which parts of the document are the best indicators of what the document is all about for future ranking. Depending on the *philosophy* of the search engine, words from the <TITLE> and <META> tags are usually scrutinized as well as the links that appear on the page. Keeping counts or frequencies of words within the entire document/webpage or just within the tags is essential for weighting the overall importance of words (see section 3.2.1).

There are many reasons why search engines automatically index. The biggest reason is time. Besides, there's no way that the search engine could copy each of the millions of documents on the Web without exhausting available storage capacity. Another critical reason for using this method is to control how new documents are added to the collection. If one tried to *index on the fly* it would be difficult to control the flow of how new documents are added to the Web. Retrieving documents and building an index provides a manageable infrastructure for information retrieval and storage.

By simply observing how current industrial search engines go about their business, search engine builders can glean two important bits of advice. One is recognizing the importance of systematically building an index and how to control additions to the text corpus. It also underscores the necessity of file preparation for the content developer and the importance of predetermining how the search engines will handle the sometimes unwieldly WWW.

2.4   Item Normalization

Building an index is more than just extracting words and building a data structure (e.g., term-by-document matrix) based on their occurrences. The words must be sliced and diced before being placed into any inverted file structure (see section 2.5). This puréeing process is referred to as *item normalization.* Kowalski in [33] summarizes it as follows:

> The first step in any integrated system is to normalize the incoming items to a standard format. In addition to translated multiple external formats that might be received into a single consistent data structure that can be manipulated by the functional processes, item normalization provides logical restructuring of the item. Additional operations during item normalization are needed to create a searchable data structure: identification of processing tokens (e.g., words), characterizations of tokens, and stemming (e.g., removing word endings) of the tokens. The original item or any of its logical subdivisions is available for the user to display. The processing tokens and their characterizations are used to define the searchable text from the total received text.

In other words, part of the document preparation is taking the smallest unit of the document, in most cases words, and constructing searchable data structures. Words are redefined as the symbols (letters, numbers) between interword symbols such as blanks. A searching system must make decisions on how to handle words, numbers, and punctuation. Documents are not just made up of words: they are composed of *processing tokens*. Identifying processing tokens constitutes the first part of item normalization. The characterization of tokens or *disambiguation* of terms (i.e., deriving the meaning of a word based on context) can be handled after normalization is complete.

A good illustration of the effort and resources required to characterize tokens automatically can be found in the National Library of Medicine's Unified Medical Language System (UMLS) project. For over a decade, the UMLS has been working on enabling computer systems to *understand* medical meaning [44]. The *Metathesaurus* is one of the components of UMLS and contains half a million biomedical concepts with over a million different concept names. Obviously, to do this, automated processing of the machine-readable versions of its 40 source vocabularies is necessary, but it also requires review and editing by subject experts.

The next step in item normalization is to apply stop lists to the collection of processing tokens. Stop lists are lists of words that have little or no value as a search term. A good example of a stop list is the list of stop words from the SMART system at Cornell University (see `ftp://ftp.cs.cornell.edu/`

`pub/smart/english.stop`). Just a quick scroll through this list of words (able, about, after, allow, became, been, before, certainly, clearly, enough, everywhere, etc.) reveals their limited impact in discriminating concepts or potential search topics. From the data compression viewpoint, stop lists eliminate the need to handle unnecessary words and reduce the size of the index and the amount of time and space required to build searchable data structures.

However, the value of removing stop words for a compressed inverted file is questionable [63]. Applying a stop list does reduce the size of the index, but the words that are omitted are typically those that require the fewest bits per pointer to store so that the overall savings in storage is not that impressive.

Although there is little debate over eliminating common words, there is some discussion on what to do about *singletons* or words that only appear once or very infrequently in a document or a collection. Some indexers may feel that the probability of searching with such a term is small or the importance of a such a word is so minimal that it should be included in the stop list also.

Stemming, or the removing of suffixes (and sometimes prefixes) to reduce a word to its root form, has a relatively long tradition in the index building process. For example, words from a document database such as *reformation, reformative, reformatory, reformed,* and *reformism* can all be stemmed to the root word *reform* (perhaps a little dangerous to remove the *re-* prefix). All five words would map to the word *reform* in the index. This saves the space of four words in the index. However, if a user queries for information about the *Reformation* and some of the returned documents describe *reformatories* (i.e., reform schools), it could leave the user scratching his or her head wondering about the quality of the search engine. If the query from the user is stemmed, there are advantages and disadvantages also. Stemming would help the user if the query was misspelled and stemming handles the plurals and common suffixes, but again there is always the risk that stemming will cause more nonrelevant items to be pulled more readily from the database. Also stemming proper nouns, such as the words that originate from database fields such as *author,* is usually not done.

Stemming can be done in various ways, but it's not a task to be regarded lightly. Stemming can be a tedious undertaking, especially considering that decisions must be made and rules developed for thousands of words in the English language. Fortunately, several automatic stemmers utilizing different approaches have been developed. As suggested by [32, 33, 63], the Porter Stemmer is one of the industry stalwarts. A public domain version (written in C) is available for downloading at `http://gatekeeper.dec.com/pub/net/` `infosys/wais/ir-book-sources/stemmer/`. The Porter Stemmer approach is based on the sequences of vowels and consonants, whereas other approaches follow the general principle of *looking up* the stem in a dictionary and assigning the stem that best represents the word.

2.5 Inverted File Structures

One of the universal linchpins of all information retrieval and database systems is the inverted file structure (IFS), a series of three components that track which documents contain which index terms. Inverted file structures provide a critical shortcut in the search process. Instead of searching the entire document database for specific terms in a query, the IFS organizes the information into an abbreviated list of terms, which then, depending on the term, references a specific set of documents. It's just like picking up a geography reference book and looking for facts about the *Appalachian Mountains.* You can turn page by page and eventually you will find your facts or you can check the index, which immediately directs you to the pages that have facts about the *Appalachian Mountains.* Both methods work, but the latter is usually much quicker.

As mentioned earlier, there are three components in the IFS:

- The *Document File* (DF) is where each document is given a unique number identifier and all the terms (processing tokens) within the document are identified.

- The *Dictionary* is a sorted list of all the unique terms (processing tokens) in the collection along with pointers to the *Inversion List.*

- The *Inversion List* (IL) contains the pointer from the term to which documents contain that term. (In a book index, the *pointer* would be the page number where the term *Appalachian* would be found.)

To illustrate the use of the IFS, we have created a two-stanza limerick about a searcher and her query.

Bread Search

There once was a searcher named Hanna,
Who needed some info on manna.
She put "rye" and "wheat" in her query
Along with "potato" or "cranbeery,"
But no mention of "sourdough" or "banana."

Instead of rye, cranberry, or wheat,
The results had more spiritual meat.
So Hanna was not pleased,
Nor was her hunger eased,
'Cause she was looking for something to eat.

2.5.1   Document File

The first step in creating the IFS is extracting the terms that should be used in the index and assigning each document a unique number. For simplicity's sake each line of the limerick can be used to represent a document (see Table 2.1). Not only is the punctuation removed, but keep in mind that common, often-used words have little value in searching and thus reside in the stop lists and are not pulled for the index. This means that a significant percentage of the words are not indexed, reducing the storage space requirements for the system. Even so, for purposes of this example we were liberal in selecting terms to be used in the index (see Table 2.2).

2.5.2   Dictionary List

The second step would be to extract the terms and create a searchable dictionary of terms. To facilitate searching, the terms could be arranged alphabetically; however, there are other time and storage saving strategies that can be implemented. Instead of whole words, the processing tokens are broken down to a letter-by-letter *molecular level* for specific data structure implementations, that is, a reorganization of the data in such a way to allow more efficient searching. Two well-known data structures for processing dictionaries are N-grams and PAT trees. Kowalski [33] gives an overview of both methods and includes references for more understanding, but for the purposes of explaining IFSs, it's unnecessary to go into (more) detail.

Table 2.1: Documents (records) defined by limerick.

1	There once was a searcher named Hanna
2	Who needed some info on manna
3	She put rye and wheat in her query
4	Along with potato or cranbeery
5	But no mention of sourdough or banana
6	Instead of rye cranberry or wheat
7	The results had more spiritual meat
8	So Hanna was not pleased
9	Nor was her hunger eased
10	Cause she was looking for something to eat

Sometimes the dictionary list might also indicate the number of times the term appears in the document database such as the list for the *Bread Search* document collection created in Table 2.3.

2.5.3 Inversion List (IL)

The final step in building an inverted file structure is to combine the dictionary list and the document list to form what is called the *inversion list*. The IL specifically *points* to a specific document(s) when a term is selected. For the limerick example, if the query contained the term *wheat*, documents (lines) 3 and 6 will be retrieved. In addition to pointing to a specific document, inverted lists can be constructed to *point* to a particular *zone* or section of the document where the term is used. Table 2.4 illustrates how both the document and position of a term can be recorded in the inverted list. Notice that the first occurrence of the word *wheat* is in document 3 at position 5 (i.e., it is the fifth word).

Inverted lists can certainly be more sophisticated than what has been described thus far, especially when the search engine must support contiguous word phrases. A *contiguous word phrase* is used when specific word combinations are requested by the user. Using the frustrated searcher Hanna, let's say she queried on *banana bread*. In other words, she wants to pull only documents where the word *banana* is next to the word *bread*. If the

Table 2.2: Terms extracted (parsed) from limerick.

Doc. No.	Terms/Key Words
1	searcher, Hanna
2	manna
3	rye, wheat, query
4	potato, cranbeery[a]
5	sourdough, banana
6	rye, cranberry,[a] wheat
7	spiritual, meat
8	Hanna
9	hunger
10	*No terms*

[a]The words *cranberry* and *cranbeery* would probably stem to the same root word, *cranb*. This illustrates the value of stemming if a word is misspelled or the writer of the document — in this case the poet — has taken *too much license.*

inverted list stores the position of each word, then such a determination can be made. Some systems are designed to measure how close terms are to each other, i.e., their *proximity*. For example, a user might want documents that only contain *banana* if it is within five words of *bread*. Obviously, it's a useful feature and many systems can do this, but there is a slight trade-off (though) in systems that support contiguous word phrases and proximity measures, higher storage requirements and computational costs.

Also there is no hard and fast rule that only one inverted file system can be created for a document collection. Separate inverted file structures can be developed for different zones or portions of the documents such as title or abstract. An inverted file system could be built for just authors with a special set of rules such as no stop lists. This allows the user to search more

Table 2.3: Dictionary list for the limerick.

Term	Global Frequency
banana	1
cranb	2
Hanna	2
hunger	1
manna	1
meat	1
potato	1
query	1
rye	2
sourdough	1
spiritual	1
wheat	2

quickly on specific fields within the database.

> Besides serving as an example of an IFS, and reinforcing the adage that life does imitate art (or in this case life imitates bad poetry), *Bread Search* also illustrates the blur between searching for concepts and more concrete items. In *Bread Search*, Hanna could have been searching a biblical database about manna, the breadlike food provided by God to the Israelites during their exodus from Egypt. But *manna* has conceptual connotations as well, mainly as a metaphor of something that can be spiritually nourishing. Perhaps in a future search in the same biblical database, Hanna could be looking for activities that give a person spiritual nourishment and regeneration—the bread of life. Hopefully, Hanna's results will include more than just recipes.

2.5.4 Other File Structures

While the IFS is commonly used to build an index, there are other approaches. One alternative is the use of *signature files* in which words/terms are converted into binary strings (composed of zeros and ones). Words from

Table 2.4: Inversion list for the limerick example.

Term	(Doc. No., Position)
banana	(5,7)
cranb	(4,5); (6,4)
Hanna	(1,7); (8,2)
hunger	(9,4)
manna	(2,6)
meat	(7,6)
potato	(4,3)
query	(3,8)
rye	(3,3); (6,3)
sourdough	(5,5)
spiritual	(7,5)
wheat	(3,5); (6,6)

a query are mapped to their signatures and the search involves matching bit positions with the (precomputed) signatures of the items/documents. In one sense, the *signature file* takes an opposite approach compared with the IFS. Whereas the IFS matches the query with the term, the signature file eliminates all nonmatches. After superimposing different signature files, the query signature is compared and the nonmatches fall by the wayside. In general, the documents corresponding to the remaining signatures are then searched to see if the query terms do indeed exist in those documents.

Figure 2.2 illustrates how one of the verses (ignoring punctuation) from the limerick in section 2.5 might be encoded as a *block* signature. Here, the first three characters of each word/term in the verse are translated into 8-bit strings according to the hash function $f(c) = 2^{(c \bmod 8)}$, where the value c is the octal value of the corresponding ASCII character. The four word signatures are *OR*'ed to create the block signature for the verse (document). To avoid signatures that are overly dense with 1's, a maximum number of words per block is typically specified and documents are partitioned into blocks of that size [33]. Also, a maximum number (m) of 1's that may be specified per word is defined ($m = 3$ in Figure 2.2). Query terms are mapped to their respective signatures and then bit positions in all signatures (query and documents) are compared to delineate nonmatches. A technique

Term	Octal Values per Character (c)		
Nor	116	157	162
her	150	145	162
hun*ger*	150	165	156
eas*ed*	145	141	163

Term	$f(c) = 2^{(c \bmod 8)}$		
Nor	01 000 000	10 000 000	00 000 100
her	00 000 001	00 100 000	00 000 100
hun*ger*	00 000 001	00 100 000	00 100 000
eas*ed*	00 100 000	00 000 010	00 001 000
Block Signature	01 100 001	10 100 010	00 101 100

Figure 2.2: Block signature construction.

referred to as *Huffman coding* is a well-documented approach for encoding symbols/words, given a certain probability distribution for the symbols [63].

Although the binary strings can become fairly large, blocking can be used to combine signatures and thus facilitate searching. One variation of the signature file is the *bitmap*. For a particular document, every bit associated with a term used in that document is set to one. All other bits are set to zero. For lengthy documents, very long binary strings will be produced, and the exhaustive storage requirements for large document collections can render this approach impractical.

According to [63] signature files require more space than compressed inverted files and are typically designed to handle large conventional databases. A more in-depth discussion of the intricacies and specific usefulness of signature files can be found in [22].

Chapter 3

Vector Space Models

As first mentioned in section 1.3, a vector space model can be used to encode/represent both terms and documents in a text collection. In such a model, each component of a document vector can be used to represent a particular term/key word, phrase, or concept used in that document. Normally, the value assigned to a component reflects the importance of the term, phrase, or concept in representing the *semantics* of the document (see section 3.2.1).

3.1 Construction

A document collection composed of n documents that are indexed by m terms can be represented as an $m \times n$ *term-by-document matrix A*. The n (column) vectors representing the n documents form the columns of the matrix. Thus, the matrix element a_{ij} is the weighted frequency at which term i occurs in document j [7]. Using the vector space model, the columns of A are interpreted as the *document vectors*, and the rows of A are considered the *term vectors*.

The column space of A essentially determines the semantic content of the collection; i.e., the document vectors span the content. However, it is not the case that every vector represented in the column space of A has a specific interpretation. For example, a linear combination of any two document vectors does not necessarily represent a viable document from the collection. More importantly, the vector space model can exploit geometric relationships between document (and term) vectors in order to explain both similarities and differences in concepts.

31

3.1.1   Term-by-Document Matrices

For heterogeneous text collections, i.e., those representing many different contexts (or topics) such as newspapers and encyclopedias, the number of terms (m) is typically much greater than the number of documents $(n \ll m)$. For the WWW, however, the situation is reversed. Specifically, a term-by-document matrix using the words of the largest English language dictionary as terms and the set of all webpages as documents would be about $300,000 \times 300,000,000$ [3, 12, 36]. Since any one document will consist of only a small subset of words from the entire dictionary (associated with the particular database), the majority of elements defining the term-by-document matrix will be zero.

Figure 3.1 demonstrates how a 9×7 term-by-document matrix is constructed from a small collection of book titles.[2] In this simple example, only a subset (underlined) of all the words used in the 7 titles were chosen terms/key words for indexing purposes. The stop list (see section 2.4) in this case would contain words like *First, Aid, Room,* etc. Notice that any particular term occurs only once in any given document. Certainly for larger collections, the term frequencies can be considerably larger than 1 for any document. As the semantic content of a document is generally determined by the *relative* frequencies of terms, the elements of the term-by-document matrix A are sometimes scaled so that the Euclidean norm (or vector 2-norm) of each column is 1. Recall that the Euclidean vector norm $\| x \|_2$ is defined by

$$\| x \|_2 = \sqrt{x^T x} = \sqrt{\sum_{i=1}^{m} x_i^2},$$

where $x = (x_1, x_2, \ldots, x_m)$. For example, with each column a_j of the matrix A in Figure 3.1 we have $\| a_j \|_2 = 1$, $j = 1, \ldots, 7$. As will be discussed in section 3.2.1, the actual values assigned to the elements of the term-by-document matrix $A = [a_{ij}]$ are usually *weighted* frequencies as opposed to the *raw* counts of term occurrences (within a document or across the entire collection).

As illustrated in Figure 3.1, not all words are used to describe the collection of book titles. Only those words related to *child safety* were selected. Determining which words to index and which words to discard defines both the *art* and the *science* of automated indexing. Using lexical matching, no

---

[2]These *actual* book titles were obtained using the search option at www.amazon.com.

	Terms		Documents
T1:	Bab(y,ies,y's)	D1:	<u>Infant</u> & <u>Toddler</u> First Aid
T2:	Child(ren's)	D2:	<u>Babies</u> & <u>Children's</u> Room (For Your <u>Home</u>)
T3:	Guide	D3:	<u>Child</u> Safety at <u>Home</u>
T4:	Health	D4:	Your <u>Baby's</u> <u>Health</u> and <u>Safety</u>: From <u>Infant</u>
T5:	Home		to <u>Toddler</u>
T6:	Infant	D5:	<u>Baby</u> <u>Proofing</u> Basics
T7:	Proofing	D6:	Your <u>Guide</u> to Easy Rust <u>Proofing</u>
T8:	Safety	D7:	Beanie <u>Babies</u> Collector's <u>Guide</u>
T9:	Toddler		

The 9×7 term-by-document matrix before normalization, where the element $\hat{a}_{ij}$ is the number of times term i appears in document title j:

$$
\hat{A} = \begin{pmatrix}
0 & 1 & 0 & 1 & 1 & 0 & 1 \\
0 & 1 & 1 & 0 & 0 & 0 & 0 \\
0 & 0 & 0 & 0 & 0 & 1 & 1 \\
0 & 0 & 0 & 1 & 0 & 0 & 0 \\
0 & 1 & 1 & 0 & 0 & 0 & 0 \\
1 & 0 & 0 & 1 & 0 & 0 & 0 \\
0 & 0 & 0 & 0 & 1 & 1 & 0 \\
0 & 0 & 1 & 1 & 0 & 0 & 0 \\
1 & 0 & 0 & 1 & 0 & 0 & 0
\end{pmatrix}
$$

The 9×7 term-by-document matrix with unit columns:

$$
A = \begin{pmatrix}
0 & 0.5774 & 0 & 0.4472 & 0.7071 & 0 & 0.7071 \\
0 & 0.5774 & 0.5774 & 0 & 0 & 0 & 0 \\
0 & 0 & 0 & 0 & 0 & 0.7071 & 0.7071 \\
0 & 0 & 0 & 0.4472 & 0 & 0 & 0 \\
0 & 0.5774 & 0.5774 & 0 & 0 & 0 & 0 \\
0.7071 & 0 & 0 & 0.4472 & 0 & 0 & 0 \\
0 & 0 & 0 & 0 & 0.7071 & 0.7071 & 0 \\
0 & 0 & 0.5774 & 0.4472 & 0 & 0 & 0 \\
0.7071 & 0 & 0 & 0.4472 & 0 & 0 & 0
\end{pmatrix}
$$

Figure 3.1: The construction of a term-by-document matrix A.

title would be retrieved for a user searching for titles on *First Aid*. Both the words *First* and *Aid* would have to be added to the index (i.e., create two new rows for the term-by-document matrix) in order to retrieve book title $D1$.

In constructing a term-by-document matrix, terms are usually identified by their word stems (see section 2.4). In the example shown in Figure 3.1, the words *Baby*, *Babies*, and *Baby's* are counted as 1 term, and the words *Child* and *Children* are treated the same. Stemming, in these situations, reduces the number of rows in the term-by-document matrix A from 12 to 9. The reduction of storage (via stemming) is certainly an important consideration for large collections of documents.

Even with this tiny sample of book titles, we find evidence of two of the most common (and major) obstacles to the retrieval of relevant information: *synonymy* and *polysemy*. Synonymy refers to the use of synonyms or different words that have the same meaning, and polysemy refers to words that have different meanings when used in varying contexts. Four of the nine terms indexed in Figure 3.1 are synonyms: *Baby*, *Child*, *Infant*, and *Toddler*. Examples of *polysemous* words include *Proofing* (child or rust) and *Babies* (human or stuffed). Methods for handling the effects of synonymy and polysemy in the context of vector space models are considered in section 3.2.

3.1.2   Simple Query Matching

The small collection of book titles from Figure 3.1 can be used to illustrate simple query matching in a low-dimensional space. Since there are exactly 9 terms used to index the 7 book titles, queries can be represented as 9×1 vectors in the same way that each of the 7 titles is represented as a column of the 9×7 term-by-document matrix A. In order to retrieve books on *Child Proofing* from this small collection, the corresponding query vector would be

$$q = (0 \quad 1 \quad 0 \quad 0 \quad 0 \quad 0 \quad 1 \quad 0 \quad 0)^T;$$

that is, the frequencies of the terms *Child* and *Proofing* in the query would specify the values of the appropriate nonzero entries in the query vector.

Query matching in the vector space model can be viewed as a search in the column space of the matrix A (i.e., the subspace *spanned* by the document vectors) for the documents most similar to the query. One of the most common similarity measures used for query matching in this context is the *cosine* of the angle between the query vector and the document vectors.

If we define a_j as the jth document vector (or the jth column of the term-by-document matrix A), then the cosines between the query vector $q = (q_1, q_2, \ldots, q_m)^T$ and the $n = 7$ document vectors are defined by

$$\cos \theta_j = \frac{a_j^T q}{\| a_j \|_2 \| q \|_2} = \frac{\sum_{i=1}^{m} a_{ij} q_i}{\sqrt{\sum_{i=1}^{m} a_{ij}^2} \sqrt{\sum_{i=1}^{m} q_i^2}} \tag{3.1}$$

for $j = 1, \ldots, n$. Since the query vector and document vectors are typically *sparse* (i.e., have relatively few nonzero elements), computing the inner products and norms in (3.1) is not that expensive. Also, notice that the document vector norms $\| a_j \|_2$ can be precomputed and stored before any cosine computation. If the query and document vectors are normalized (see section 3.2.1) so that $\| q \|_2 = \| a_j \|_2 = 1$, the cosine calculation constitutes a single inner product. More information on alternative similarity measures is provided in [30, 61].

In practice [7], documents whose corresponding document vectors produce cosines (with the query vector) greater than some threshold value (e.g., $|\cos \theta_j| \geq 0.5$) are judged as relevant to the user's query. For the collection of book titles in Figure 3.1, the nonzero cosines are $\cos \theta_2 = \cos \theta_3 = 0.4082$, and $\cos \theta_5 = \cos \theta_6 = 0.5000$. Hence, a cosine threshold of 0.5 would judge only the fifth and sixth documents as relevant to *Child Proofing*. While the fifth document is certainly relevant to the query, clearly the sixth document (concerning *rust* proofing) is irrelevant. While the seventh document is correctly ignored, the first four documents would not be returned as relevant.

If one was interested in finding books on *Child Home Safety* from the small collection in Figure 3.1, the only nonzero cosines made with the query vector

$$\tilde{q} = (0 \quad 1 \quad 0 \quad 0 \quad 1 \quad 0 \quad 0 \quad 1 \quad 0)^T$$

would be $\cos \theta_2 = 0.6667$, $\cos \theta_3 = 1.0000$, and $\cos \theta_4 = 0.2582$. With a cosine threshold of 0.5, the first, fourth, and fifth documents (which are relevant to the query) would not be returned.

Certainly, this representation of documents solely based on term frequencies does not adequately model the semantic content of the book titles. Techniques to improve the vector space representation of documents have been developed to address the *errors* or uncertainty associated with this *basic* vector space IR model. One approach is based on the use of term weights (see section 3.2.1), and another approach relies on computing low-rank approximations to the original term-by-document matrix. The premise of the latter approach is based on the potential *noise* reduction (due to problems like synonymy and polysemy) achieved by low-rank approximation. Vector space IR models such as LSI [6, 7, 18] rely on such approximations to encode both terms and documents for conceptual-based query matching. Following [6], the process of *rank reduction* can be easily explained using numerical algorithms such as the QR factorization and SVD [26], which are typically presented in linear algebra textbooks. We will discuss these important numerical methods in the context of IR modeling in chapter 4.

3.2   Design Issues

3.2.1   Term Weighting

A collection of n documents indexed by m terms (or key words) can be represented as an $m \times n$ term-by-document matrix $A = [a_{ij}]$ (see section 3.1.1). Each element, a_{ij}, of the matrix A is usually defined to be a *weighted* frequency at which term i occurs in document j [7, 49]. The main purpose for term weighting is to improve retrieval performance. Performance in this case refers to the ability to retrieve relevant information (recall) and to dismiss irrelevant information (precision). As will be discussed later in section 6.1, *recall* is measured as the ratio of the number of relevant documents retrieved to the total number of relevant items that exist in the collection, and *precision* is measured as the ratio of the number of relevant documents retrieved to the total number of documents retrieved. A desirable IR system is one that achieves high precision for most levels of recall (if not all). One way to improve recall for a given query is to use words with high frequency, i.e., those that appear in many documents in the collection. In contrast, obtaining high precision may require the use of very specific terms or words that will match the most relevant documents in the collection. No doubt some sort of compromise must be made to achieve sufficient recall without poor precision. Term weighting is one approach commonly used to improve the retrieval performance of automatic indexing systems.

Using a format similar to that presented in [31], let each element a_{ij} be

Table 3.1: Formulas for local term weights (l_{ij}).

Symbol	Name	Formula
b	Binary [49]	$\chi(f_{ij})$
l	Logarithmic [27]	$\log(1 + f_{ij})$
n	Augmented normalized Term frequency [27, 49]	$(\chi(f_{ij}) + (f_{ij}/\max_k f_{kj}))/2$
t	Term frequency [49]	f_{ij}

defined by

$$a_{ij} = l_{ij} g_i d_j, \tag{3.2}$$

where l_{ij} is the local weight for term i occurring in document j, g_i is the global weight for term i in the collection, and d_j is a document normalization factor that specifies whether or not the columns of A (i.e., the documents) are normalized. Tables 3.1 through 3.3 contain some of the popular weight formulas used in automated indexing systems. For convenience [31], let

$$\chi(r) = \begin{cases} 1 & \text{if } r > 0, \\ 0 & \text{if } r = 0, \end{cases}$$

define f_{ij} as the number of times (frequency) that term i appears in document j, and let $p_{ij} = f_{ij}/\sum_j f_{ij}$.

A simple notation for specifying a term weighting scheme is to use the three-letter string associated with the particular local, global, and normalization factors desired. For example, the *lfc* weighting scheme defines

$$a_{ij} = \frac{\log(f_{ij} + 1)\log(n/\sum_j \chi(f_{ij}))}{\sqrt{\sum_i \left(\log(f_{ij} + 1)\log(n/\sum_j \chi(f_{ij}))\right)^2}}.$$

Defining an appropriate weighting scheme from the choices in Tables 3.1 through 3.3 certainly depends on certain characteristics of the document collection [49]. The choice for the local weight (l_{ij}) may well depend on the

Table 3.2: Formulas for global term weights (g_i).

Symbol	Name	Formula
x	None	1
e	Entropy [20]	$1 + \left(\sum_j (p_{ij}\log(p_{ij}))/\log n \right)$
f	Inverse document frequency (IDF) [20, 49]	$\log\left(n/\sum_j \chi(f_{ij}) \right)$
g	GfIdf [20]	$\left(\sum_j f_{ij} \right)/\sum_j \chi(f_{ij})$
n	Normal [20]	$1/\sqrt{\sum_j f_{ij}^2}$
p	Probabilistic Inverse [27, 49]	$\log\left((n - \sum_j \chi(f_{ij}))/\sum_j \chi(f_{ij}) \right)$

vocabulary or word usage patterns for the collection. For technical or scientific vocabularies (e.g., technical reports and journal articles), schemes of the form *nxx* with normalized term frequencies are generally recommended. For more general (or varied) vocabularies (e.g., popular magazines, encyclopedias), simple term frequencies (*t***) may be sufficient. Binary term frequencies (*b***) are useful when the term list (or row dimension of the term-by-document matrix) is relatively short, such as the case with *controlled vocabularies*.

Choosing the global weighting factor (g_i) should take into account the state of the document collection. By this we mean how often the collection is likely to change. Adjusting the global weights in response to new vocabulary will impact all the corresponding rows of the term-by-document matrix. To avoid updating, one may simply disregard the global factor (**x**) altogether. For more static collections, the inverse document frequency (IDF) global weight (**f**) is a common choice among automatic indexing systems [20, 49].

It has been observed that the probability that a document being judged relevant by a user significantly increases with document length [57]. In other words, the longer a document is, the more likely all the key words will be found (and with higher frequency). As demonstrated for SMART [16],

Table 3.3: Formulas for document normalization (d_j).

Symbol	Name	Formula
x	None	1
c	Cosine [49]	$\left(\sum_i (g_i l_{ij})^2 \right)^{-1/2}$

experiments with document length normalization have demonstrated that the traditional cosine normalization ($**c$) is not particularly effective for large full text documents (e.g., TREC-4). In order to retrieve documents of a certain length with the same probability of retrieving a relevant document of that same length, the recent *Lnu* or *pivoted-cosine* normalization scheme [16, 57] has been proposed for indexing the TREC collections. This scheme is based on the assignment

$$a_{ij} = \frac{(1 + \log(fij \rightarrow fij))/(1 + \log(\bar{f}_{ij}))}{(1 - s) \times p + s \times u},$$

where $\bar{f}_{ij} = \left(\sum_i f_{ij} \right) / \left(\sum_i \chi(f_{ij}) \right)$, s is referred to as the *slope* and is typically set to 0.2, p is the *pivot* value that is defined to be the average number of *unique* terms/key words occurring (per document) throughout the collection, and u is the number of unique terms in document j. In effect, this formula is an adjustment to cosine normalization so that relevant documents of smaller size will have a better chance of being judged similar. In TREC-3 experiments [57], the *Lnu* weighting scheme employing *pivoted-cosine* normalization obtained 13.7% more relevant documents (for 50 queries) compared with a comparable weighting scheme based on the more traditional cosine normalization. What this result indicates is that document relevance and length should not be considered mutually independent with respect to retrieval performance.

3.2.2 Sparse Matrix Storage

As discussed in sections 1.3 and 3.1.1, the number of nonzeros defined within term-by-document matrices is relatively small compared with the number of

zeros. Such sparse matrices generally lack any particular nonzero structure
or pattern such as *banded* 10×6 matrix A illustrated by

$$
A = \begin{pmatrix}
x & x & & & & \\
x & x & & & & \\
 & x & x & & & \\
 & & x & x & & \\
 & & x & x & & \\
 & & & x & x & \\
 & & & x & x & \\
 & & & x & x & \\
 & & & & x & x \\
 & & & & x & x
\end{pmatrix} .
$$

If a term-by-document matrix had a (banded) nonzero element structure
similar to that above, the ability to identify *clusters* of documents sharing
similar terms would be quite simplified. Obtaining such matrices for general
text is quite difficult; however, some progress in the reordering of hypertext-
based matrices has been made [8].

In order to avoid both the storage and processing of zero elements, a variety of sparse matrix storage formats have been developed [2]. In order to
allocate contiguous storage locations in memory for the nonzero elements,
a sparse matrix format must *know* exactly how the elements *fit* into the
complete (or full) term-by-document matrix. Two formats that are suitable
for term-by-document matrices are the *compressed row storage* (CRS) and
compressed column storage (CCS). These sparse matrix formats do not make
any assumptions on the nonzero structure (or pattern) of the matrix. Compressed row storage places the nonzeros of the matrix rows into contiguous
storage (arrays), and the CCS format stores the matrix columns in a similar
fashion. To implement either format requires three arrays of storage that
can be used to access/store any nonzero. The contents of these arrays for
both the CRS and CCS formats are described below.

Compressed Row Storage (CRS)

This sparse matrix format requires one floating-point array (`val`) for stor-
ing the nonzero values (i.e., weighted or unweighted term frequencies), and
two additional integer arrays for indices (`col_ind`, `row_ptr`). The `val` array stores the nonzero elements of the term-by-document matrix A as they

are traversed row-wise; i.e., store all the frequencies (in order from left to right) of the current term before moving on to the next one. The `col_ind` array stores the corresponding column indices (document numbers) of the elements (term frequencies) in the `val` array. Hence, `val(k)` $= a_{ij}$ implies `col_ind(k)` $= j$. The `row_ptr` array stores the locations of the `val` array (term frequencies) that begin a row. If `val(k)` $= a_{ij}$, then `row_ptr(i)` $\leq$ `k` $\leq$ `row_ptr(i+1)`. If nnz is the number of nonzero elements (term frequencies) for the term-by-document matrix A, then it is customary to define `row_ptr(n+1)` $= nnz+1$. It is easy to show that the difference `row_ptr(i+1)` $-$ `row_ptr(i)` indicates how many nonzeros are in the ith row of the matrix A, i.e., the number of documents associated with the ith term/key word. For an $m \times n$ term-by-document matrix, the CRS format requires only $2nnz + m + 1$ storage (array) locations compared with mn for the complete (includes zeros) matrix A.

Compressed Column Storage (CCS)

The CCS format is almost identical to the CRS format in that the columns of the matrix A (as opposed to the rows) are stored in contiguous (array) locations. The CCS format, which is the CRS format for the transpose of the matrix A (i.e., A^T), is also known as the *Harwell–Boeing* sparse matrix format (see [19]). The three arrays required for the CCS format are {`val`, `row_ind`, `col_ptr`}, where `val` stores the the nonzero elements of the term-by-document matrix A as they are traversed columnwise, i.e., stores all the frequencies (in order from top to bottom) of each document. The `row_ind` array stores the corresponding row indices (term numbers or identifiers) of the elements (term frequencies) in the `val` array. Hence, `val(k)` $= a_{ij}$ implies `row_ind(k)` $= i$. The `col_ptr` array with CCS stores the locations of the `val` array (term frequencies) that begin a column (document) so that `val(k)` $= a_{ij}$ indicates that `col_ptr(j)` $\leq$ `k` $\leq$ `col_ptr(j+1)`. Figure 3.2 illustrates the 3 arrays needed for the CRS and CCS representations of the 9×7 term-by-document matrix A in Figure 3.1.

3.2.3 Low-Rank Approximations

The process of indexing (whether manual or automated) may well be considered an art rather than a science. The uncertainties associated with term-by-document matrices can largely be attributed to differences in lan-

CRS			CCS		
val	col_ind	row_ptr	val	row_ind	col_ptr
0.5774	2	1	0.7071	6	1
0.4472	4	5	0.7071	9	3
0.7071	5	7	0.5774	1	6
0.7071	7	9	0.5774	2	9
0.5774	2	10	0.5774	5	14
0.5774	3	12	0.5774	2	16
0.7071	6	14	0.5774	5	18
0.7071	7	16	0.5774	8	20
0.4472	4	18	0.4472	1	
0.5774	2	20	0.4472	4	
0.5774	3		0.4472	6	
0.7071	1		0.4472	8	
0.4472	4		0.4472	9	
0.7071	5		0.7071	1	
0.7071	6		0.7071	7	
0.5774	3		0.7071	3	
0.4472	4		0.7071	7	
0.7071	1		0.7071	1	
0.4472	4		0.7071	3	

Figure 3.2: CRS and CCS representations of the 9×7 term-by-document matrix A ($nnz = 19$) from Figure 3.1.

guage (word choice) and culture. Errors in measurement can accumulate and thereby generate *uncertainty* in the experimental data. Fundamental differences in word usage between authors and readers suggest that there will never be a *perfect* term-by-document matrix that accurately represents all possible term-document associations. As an example, notice that document D4: Your Baby's Health and Safety: From Infant to Toddler from the small collection of booktitles in Figure 3.1 would be a good match for a search on books for *child proofing*. This judgment of *relevance* would suggest that the unnormalized matrix $\hat{A}$ in Figure 3.1 should have the entries $\hat{a}_{24} = \hat{a}_{74} = 1$. Since the true association of terms (and concepts) to documents is subject to many interpretations, the term-by-document matrix A may be better represented [6] by the matrix sum $A + E$, where E reflects the *error* or *uncertainty* in assigning (or generating) the elements of matrix A.

It can be shown that the 9×7 term-by-document matrix for the small collection of booktitles in Figure 3.1 has rank 7. In other words, the matrix has exactly 7 linearly independent columns (or document vectors in this case) [26]. For larger text collections and especially the WWW, the corresponding $m \times n$ matrix A may not have *full* rank (i.e., n linearly independent columns). In fact, if the booktitle

D8: Safety Guide for Child Proofing Your Home

was added to the collection in Figure 3.1, the unnormalized matrix $\hat{A}$ would still have rank 7. In other words, both the matrix $\hat{A}$ given by

$$\hat{A} = \begin{pmatrix} 0 & 1 & 0 & 1 & 1 & 0 & 1 & 0 \\ 0 & 1 & 1 & 0 & 0 & 0 & 0 & 1 \\ 0 & 0 & 0 & 0 & 0 & 1 & 1 & 1 \\ 0 & 0 & 0 & 1 & 0 & 0 & 0 & 0 \\ 0 & 1 & 1 & 0 & 0 & 0 & 0 & 1 \\ 1 & 0 & 0 & 1 & 0 & 0 & 0 & 1 \\ 0 & 0 & 1 & 1 & 0 & 0 & 0 & 1 \\ 1 & 0 & 0 & 1 & 0 & 0 & 0 & 0 \end{pmatrix}$$

and the normalized term-by-document matrix A defined as

$$A = \begin{pmatrix} 0 & 0.5774 & 0 & 0.4472 & 0.7071 & 0 & 0.7071 & 0 \\ 0 & 0.5774 & 0.5774 & 0 & 0 & 0 & 0 & 0.4472 \\ 0 & 0 & 0 & 0 & 0 & 0.7071 & 0.7071 & 0.4472 \\ 0 & 0 & 0 & 0.4472 & 0 & 0 & 0 & 0 \\ 0 & 0.5774 & 0.5774 & 0 & 0 & 0 & 0 & 0.4472 \\ 0.7071 & 0 & 0 & 0.4472 & 0 & 0 & 0 & 0 \\ 0 & 0 & 0 & 0 & 0.7071 & 0.7071 & 0 & 0.4472 \\ 0 & 0 & 0.5774 & 0.4472 & 0 & 0 & 0 & 0.4472 \\ 0.7071 & 0 & 0 & 0.4472 & 0 & 0 & 0 & 0 \end{pmatrix}$$

would still have 7 linearly independent columns (document vectors).

Recent approaches to facilitate conceptual indexing, i.e., the ability to find relevant information without requiring literal word matches, have focused on the use of rank-k approximations to term-by-document matrices [6, 11, 31]. Latent semantic indexing [7, 18] is one popular approach that uses such approximations to encode m terms and n documents in k-dimensional space, where $k \ll \min(m,n)$. As illustrated in Figure 3.3, a rank-2 approximation to the matrix A from Figure 3.1 can be used to represent both terms and documents in two dimensions. Unlike the traditional vector space model, the coordinates produced by low-rank approximations do not *explicitly* reflect term frequencies within documents. Instead, methods such as LSI attempt to model *global* usage patterns of terms so that related documents that may not share common (literal) terms are still represented by nearby vectors in a k-dimensional subspace. The grouping of terms (or documents) in the subspace serves as an automated approach for *clustering* information according to concepts [33]. From Figure 3.3, the separation of booktitles in two dimensions reflects both the isolated use of terms such as *Guide* and *Proofing* and the used synonyms such as *Baby*, *Child*, *Infant*, and *Toddler*. Notice that the vector representation of the query *Child Home Safety* is clearly in the direction of the document cluster {D1, D2, D3, D4}. In fact, the largest cosine values between the query vector and all document vectors produced by the rank-2 approximation are $\cos \theta_2 = 1.000$, $\cos \theta_4 = 0.9760$, $\cos \theta_1 = 0.9788$, and $\cos \theta_2 = 0.8716$. In contrast to the simple query matching discussed in section 3.1.2, this particular IR model would return the first and fourth documents as relevant to the query. The fifth document, however, would still be missed even with this improved encoding method. Determining the optimal rank to encode the original term-by-document matrix A is an open question [7] and is certainly database dependent. In order to understand just how the coordinates shown in Figure 3.3 are produced, we turn our attention to matrix factorizations that can be used to produce these low-rank approximations.

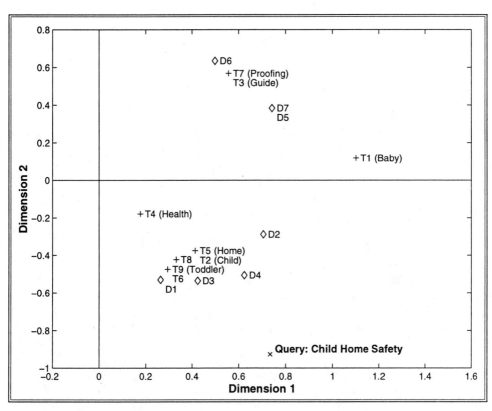

Figure 3.3: Two-dimensional representation of the booktitles collection from Figure 3.1.

Chapter 4

Matrix Decompositions

To produce a reduced-rank approximation of an $m \times n$ term-by-document matrix A, one must first be able to identify the dependence between the columns or rows of the matrix A. For a rank-k matrix A, the k basis vectors of its column space serve in place of its n column vectors to represent its column space.

4.1   QR Factorization

One set of basis vectors is found by computing the QR factorization of the term-by-document matrix

$$A = QR, \tag{4.1}$$

where Q is an $m \times m$ orthogonal matrix and R is an $m \times n$ upper triangular matrix. Recall that a square matrix Q is *orthogonal* if its columns are orthonormal. In other words, if q_j denotes a column of the orthogonal matrix Q, then q_j has unit Euclidean norm ($\| q_j \|_2 = \sqrt{q_j^T q_j} = 1$ for $j = 1, 2, \ldots, m$) and it is orthogonal to all other columns of Q ($\sqrt{q_j^T q_i} = 0$ for all $i \neq j$). The rows of Q are also orthonormal meaning that $Q^T Q = Q Q^T = I$. The factorization of the matrix A in equation (4.1) exists for any matrix A and [26] surveys the various methods for computing the QR factorization. Given the relation $A = QR$, it follows that the columns of the matrix A are all linear combinations of the columns of Q. Thus, a subset of k of the columns of Q form a *basis* for the column space of A, where $k = \mathrm{rank}(A)$.

The QR factorization in equation (4.1) can be used to determine the basis vectors for any term-by-document matrix A. For the 9×7 term-by-document matrix in Figure 3.1, the factors are

$$Q[:, 1:5] \; = \; \begin{pmatrix} 0 & -0.5774 & 0.5164 & -0.4961 & 0.1432 \\ 0 & -0.5774 & -0.2582 & 0.2481 & -0.0716 \\ 0 & 0 & 0 & 0 & 0 \\ 0 & 0 & 0 & -0.6202 & -0.2864 \\ 0 & -0.5774 & -0.2582 & 0.2481 & -0.0716 \\ -0.7071 & 0 & 0 & 0 & 0 \\ 0 & 0 & 0 & 0 & 0.9309 \\ 0 & 0 & -0.7746 & -0.4961 & 0.1432 \\ -0.7071 & 0 & 0 & 0 & 0 \end{pmatrix},$$

$$Q[:, 6:9] \; = \; \begin{pmatrix} -0.1252 & 0.3430 & 0 & 0 \\ 0.0626 & -0.1715 & -0.5962 & 0.3802 \\ 0.9393 & 0.3430 & 0 & 0 \\ 0.2505 & -0.6860 & 0 & 0 \\ 0.0626 & -0.1715 & 0.5962 & -0.3802 \\ 0 & 0 & -0.3802 & -0.5962 \\ 0.1252 & -0.3430 & 0 & 0 \\ -0.1252 & 0.3430 & 0 & 0 \\ 0 & 0 & 0.3802 & 0.5962 \end{pmatrix}, \quad (4.2)$$

$$R \; = \; \begin{pmatrix} -1 & 0 & 0 & -0.6325 & 0 & 0 & 0 \\ 0 & -1 & -0.6667 & -0.2582 & -0.4082 & 0 & -0.4082 \\ 0 & 0 & -0.7454 & -0.1155 & 0.3651 & 0 & 0.3651 \\ 0 & 0 & 0 & -0.7211 & -0.3508 & 0 & -0.3508 \\ 0 & 0 & 0 & 0 & 0.7596 & 0.6583 & 0.1013 \\ 0 & 0 & 0 & 0 & 0 & 0.7528 & 0.5756 \\ 0 & 0 & 0 & 0 & 0 & 0 & 0.4851 \\ \hline 0 & 0 & 0 & 0 & 0 & 0 & 0 \\ 0 & 0 & 0 & 0 & 0 & 0 & 0 \end{pmatrix}, (4.3)$$

where $Q[:, i:j]$ refers to columns i through j of matrix Q using Matlab [41] indexing notation.

Notice how the first 7 columns of Q are partitioned (or separated) from the others in equation (4.2). Similarly, the first 7 rows of the matrix R in equation (4.3) are partitioned from the bottom 2×7 zero submatrix. The

QR factorization of this term-by-document matrix can then be represented as

$$A = (Q_1 \quad Q_2) \begin{pmatrix} R_1 \\ 0 \end{pmatrix}$$
$$= Q_1 R_1 + Q_2 \cdot 0 = Q_1 R_1, \tag{4.4}$$

where Q_1 is the 9×7 matrix defining the first 7 columns of Q, Q_2 is the 9×2 remaining submatrix of Q, and R_1 reflects the *nonzero* rows of R. This partitioning clearly reveals that the columns of Q_2 will not contribute any nonzero values to inner products associated with the multiplication of factors Q and R to produce the matrix A. Hence, the ranks (i.e., number of independent columns) of the 3 matrices A, R, and R_1 are the same, so that the first 7 columns of Q constitute a basis for the column space of A.

As discussed in [6], the partitioning of the matrix R above into zero and nonzero submatrices is not always automatic. In many cases, *column pivoting* is needed during the QR factorization to guarantee the zero submatrix at the bottom of R (see [26] for more details).

One motivation for computing the QR factorization of the term-by-document matrix A is that the basis vectors (of the column space of A) can be used to describe the *semantic* content of the corresponding text collection. The cosines of the angles θ_j between a query vector q and the document vectors a_j (for $j = 1, 2, \ldots, n$) are given by

$$\cos \theta_j = \frac{a_j^T q}{\| a_j \|_2 \| q \|_2} = \frac{(Q_1 r_j)^T q}{\| Q_1 r_j \|_2 \| q \|_2} = \frac{r_j^T (Q_1^T q)}{\| r_j \|_2 \| q \|_2}, \tag{4.5}$$

where r_j refers to column j of the submatrix R_1. Since multiplication of a vector by a matrix having orthonormal columns does not alter the norm of the vector, we can write

$$\| Q_1 r_j \|_2 = \sqrt{(Q_1 r_j)^T Q_1 r_j} = \sqrt{r_j^T Q_1^T Q_1 r_j} = \sqrt{r_j^T I r_j} = \sqrt{r_j^T r_j} = \| r_j \|_2.$$

For the term-by-document matrix from Figure 3.1 and the query vector q (*Child Proofing*) we observe no loss of information in using the factorization in equation (4.1). Specifically, the nonzero cosines computed via equation (4.5) are identical with those computed using equation (3.1): $\cos \theta_2 = \cos \theta_3 = 0.408248$, and $\cos \theta_5 = \cos \theta_6 = 0.500000$.

Since the upper triangular matrix R and the original term-by-document matrix A have the same rank, we now focus on how the QR factorization can be used to produce a low-rank approximation (see section 3.2.3) to A. Given any term-by-document matrix A, its rank is not immediately known. The rank of the corresponding matrix R from the QR factorization of A, however, is simply the number of nonzero elements on its diagonal. With column pivoting, a permutation matrix P is generated so that $AP = QR$, and the large and small elements of the matrix R are *separated*, i.e., moving the larger entries toward the upper left corner of the matrix and the smaller ones toward the lower right. If successful, this separation essentially partitions the matrix R so that the submatrix of smallest elements is completely isolated.

With column pivoting, the matrix R in equation (4.3) is replaced by

$$
R = \left(\begin{array}{ccccc|cc}
-1 & 0 & 0 & -0.4082 & -0.4082 & -0.2582 & -0.6667 \\
0 & -1 & 0 & 0 & 0 & -0.6325 & 0 \\
0 & 0 & -1 & -0.5000 & -0.5000 & 0 & 0 \\
0 & 0 & 0 & -0.7638 & -0.1091 & -0.2760 & 0.3563 \\
0 & 0 & 0 & 0 & 0.7559 & 0.2390 & -0.3086 \\
\hline
0 & 0 & 0 & 0 & 0 & 0.6325 & 0.4082 \\
0 & 0 & 0 & 0 & 0 & 0 & -0.4082 \\
0 & 0 & 0 & 0 & 0 & 0 & 0 \\
0 & 0 & 0 & 0 & 0 & 0 & 0
\end{array}\right) \quad (4.6)
$$

$$
= \begin{pmatrix} R_{11} & R_{12} \\ 0 & R_{22} \end{pmatrix}.
$$

The submatrix R_{22} above is a relatively small part of the matrix R. In fact, $\| R_{22} \|_F / \| R \|_F = 0.8563/2.6458 = 0.3237$, where the Frobenius matrix norm ($\| \cdot \|_F$) of a real $m \times n$ matrix $B = [b_{ij}]$ is defined [26] by

$$
\|B\|_F = \sqrt{\sum_{i=1}^{m} \sum_{j=1}^{n} b_{ij}^2}.
$$

If we redefine the submatrix R_{22} to be the 4×2 zero matrix, then the modified upper triangular matrix $\tilde{R}$ has rank 5 rather than 7. The matrix $A + E = Q\tilde{R}$ also has rank 5, where $E = (A + E) - A$. The perturbation or *uncertainty* matrix E can be defined as

$$
E = (A + E) - A
$$

$$= Q \begin{pmatrix} R_{11} & R_{12} \\ 0 & 0 \end{pmatrix} - Q \begin{pmatrix} R_{11} & R_{12} \\ 0 & R_{22} \end{pmatrix}$$

$$= Q \begin{pmatrix} 0 & 0 \\ 0 & -R_{22} \end{pmatrix},$$

so that $\| E \|_F = \| R_{22} \|_F$. Since $\| E \|_F = \| R_{22} \|_F$, $\| E \|_F / \| A \|_F = \| R_{22} \|_F / \| R \|_F = 0.3237$. Hence, the relative change of about 32% in the matrix R yields the same change in the matrix A, and the rank of these two matrices is reduced from 7 to 5. As discussed in [6] and [28], studies have indicated that discrepancies in indexing (same document indexed by different professional indexers) can account for some of the uncertainty (around 20%) in IR modeling. To account for this uncertainty or *noise* in the original term-by-document matrix (A), lower-rank approximations ($A + E$) can be constructed (as shown above) so that small perturbations (E) are relatively small. Also, computing cosines via equation (4.5) requires the factorization $Q\tilde{R}$ as opposed to the (explicit) matrix $A + E$.

Returning to our small collection of booktitles in Figure 3.1, suppose we replace the original term-by-document matrix A by the perturbed matrix $A + E$ defined above. Keep in mind that the rank of $A + E = Q\tilde{R}$ is now 5 as opposed to 7. The factors Q, $\tilde{R}$, and permutation matrix P produced by column pivoting on the original term-by-document matrix A are

$$Q = \begin{pmatrix}
-0.5774 & 0 & 0 & -0.6172 & 0.5345 \\
-0.5774 & 0 & 0 & 0.3086 & -0.2673 \\
0 & 0 & -0.7071 & 0.4629 & 0.5345 \\
0 & 0 & 0 & 0 & -0.0000 \\
-0.5774 & 0 & 0 & 0.3086 & -0.2673 \\
0 & -0.7071 & 0 & 0 & -0.0000 \\
0 & 0 & -0.7071 & -0.4629 & -0.5345 \\
0 & 0 & 0 & 0 & 0 \\
0 & -0.7071 & 0 & 0 & 0
\end{pmatrix},$$

$$\tilde{R} = \begin{pmatrix}
-1 & 0 & 0 & -0.4082 & -0.4082 & -0.2582 & -0.6667 \\
0 & -1 & 0 & 0 & 0 & -0.6325 & 0 \\
0 & 0 & -1 & -0.5000 & -0.5000 & 0 & 0 \\
0 & 0 & 0 & -0.7638 & -0.1091 & -0.2760 & 0.3563 \\
0 & 0 & 0 & 0 & 0.7559 & 0.2390 & -0.3086
\end{pmatrix},$$

and

$$P = \begin{pmatrix} 0 & 1 & 0 & 0 & 0 & 0 & 0 \\ 1 & 0 & 0 & 0 & 0 & 0 & 0 \\ 0 & 0 & 0 & 0 & 0 & 0 & 1 \\ 0 & 0 & 0 & 0 & 0 & 1 & 0 \\ 0 & 0 & 0 & 1 & 0 & 0 & 0 \\ 0 & 0 & 1 & 0 & 0 & 0 & 0 \\ 0 & 0 & 0 & 0 & 1 & 0 & 0 \end{pmatrix}.$$

Using the above factors and equation (4.5), we see that the cosines with respect to the query vector q (*Child Proofing*) are very similar to those reported in section 3.1.2: $\cos\theta_2 = 0.408248$ and $\cos\theta_3 = \cos\theta_5 = \cos\theta_6 = 0.500000$. For the query vector $\tilde{q}$ (*Child Home Safety*), the rank-5 representation of $A + E$ produces only 2 nonzero cosines, $\cos\theta_2 = 0.666667$ and $\cos\theta_3 = 0.816497$, so that the previous similarity with document D4 (albeit rather small) has been lost. With a cosine threshold of 0.5 (see section 3.1.2), we would pick up the relevant document D3 for the query q but lose the relevant document D4 for query $\tilde{q}$. Hence, minor rank reduction does not necessarily improve matching for all queries.

If we wanted to further reduce the rank of R in equation (4.7), we might include both the fifth row and column in R_{22}. In this situation, we would have $\| R_{22} \|_F / \| R \|_F = 0.7559$ so that the discarding of R_{22} for a rank-4 approximation of the original term-by-document matrix A would produce a relative change of almost 76%. The nonzero cosines for the query vector q mentioned above would be $\cos\theta_2 = 0.408248$, $\cos\theta_3 = 0.308607$, $\cos\theta_4 = 0.183942$, $\cos\theta_5 = \cos\theta_6 = 0.500000$, and $\cos\theta_7 = 0.654654$. Although the relevant document D4 is now barely similar to the query (*Child Proofing*), we find that the completely unrelated document D7 is now judged as the most similar.

For the query vector $\tilde{q}$ (*Child Home Safety*), the rank-4 approximation produces the nonzero cosines: $\cos\theta_2 = 0.666667$, $\cos\theta_3 = 0.755929$, $\cos\theta_4 = 0.100125$, and $\cos\theta_7 = 0.356348$. Here again, an increase in the number of irrelevant documents (D7) matched may coincide with attempts to match documents (D4) previously missed with larger ranks. Explanations as to why one variant (rank reduction) of a term-by-document matrix works better than another for particular queries are not consistent. As further discussed in [6], it is possible to improve the performance of a vector space IR model (e.g., LSI) by reducing the rank of the term-by-document matrix A. Note that even the 32% change in A used above for the rank-5 approximation can be considered quite large in the context of scientific or engineering applications where accuracies of three or more decimal places (0.1% error or better) are needed.

4.2 Singular Value Decomposition (SVD)

In section 4.1, we demonstrated the use of the QR factorization to generate document vectors of any specific dimension. While this approach does provide a reduced-rank basis for the column space of the term-by-document matrix A, it does not provide any information about the row space of A. In this section, we introduce an alternate SVD-based approach (although more demanding from a computational standpoint), which provides reduced-rank approximations to both spaces. Furthermore, the SVD has the unique mathematical feature of providing the rank-k approximation to a matrix A of minimal change for any value of k.

The SVD of the $m \times n$ term-by-document matrix A is written

$$A = U\Sigma V^T,$$

where U is the $m \times m$ orthogonal matrix whose columns define the left singular vectors of A, V is the $n \times n$ orthogonal matrix whose columns define the right singular vectors of A, and Σ is the $m \times n$ diagonal matrix containing the singular values $\sigma_1 \geq \sigma_2 \geq \cdots \geq \sigma_{\min(m,n)}$ of A in order along its diagonal. We note that this factorization exists for any matrix A and that methods for computing the SVD of both dense [26] and sparse matrices (see chapter 9) are well documented. The relative sizes of the factors U, Σ, and V for the cases $m > n$ and $m < n$ are illustrated in Figure 4.1. All off-diagonal elements of the Σ matrix are zeros.

Both the SVD $A = U\Sigma V^T$ and the QR factorization $AP = QR$ can be used to reveal the rank (r_A) of the matrix A. Recall that r_A is the number of nonzero diagonal elements of R. Similarly, r_A is also the number of nonzero diagonal elements of Σ. Whereas the first r_A columns of Q form a basis for the column space,[3] so do the first r_A columns of U. Since a rank-k approximation to A, where $k \leq r_A$, can be constructed by ignoring

---

[3] The first r_A rows of V^T form a basis for the row space of A.

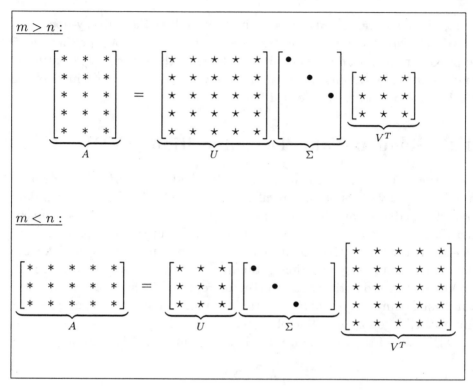

Figure 4.1: Component matrices of the SVD [6].

(or setting equal to zero) all but the first k rows of R, we can define an alternative rank-k approximation (A_k) to the matrix A by setting all but the k-largest singular values of A equal to zero. As discussed in [6, 7], this approximation is in fact the *closest* rank-k approximation to A according to a theorem by Eckart and Young in [21, 43]. This theorem demonstrated that the error in approximating A by A_k is given by

$$\|A - A_k\|_F = \min_{\text{rank}(B) \leq k} \|A - B\|_F = \sqrt{\sigma_{k+1}^2 + \cdots + \sigma_{r_A}^2}, \qquad (4.7)$$

where $A_k = U_k \Sigma_k V_k^T$, U_k and V_k compose the first k columns of U and V, respectively, and Σ_k is the $k \times k$ diagonal matrix containing the k-largest singular values of A. In other words, the error in approximating the original term-by-document matrix A by A_k is determined by the truncated (or discarded) singular values (σ_{k+1}, σ_{k+2}, $\ldots$, σ_{r_A}).

The SVD of the matrix A in Figure 3.1 is $A = U\Sigma V^T$, where

$$U[:, 1:5] = \begin{pmatrix}
0.6977 & 0.0931 & -0.0175 & 0.6951 & 0 \\
0.2619 & -0.2966 & -0.4681 & -0.1969 & 0 \\
0.3527 & 0.4491 & 0.1017 & -0.4013 & -0.7071 \\
0.1121 & -0.1410 & 0.1478 & 0.0733 & -0.0000 \\
0.2619 & -0.2966 & -0.4681 & -0.1969 & 0 \\
0.1874 & -0.3747 & 0.5049 & -0.1270 & 0 \\
0.3527 & 0.4491 & 0.1017 & -0.4013 & 0.7071 \\
0.2104 & -0.3337 & -0.0954 & -0.2820 & 0 \\
0.1874 & -0.3747 & 0.5049 & -0.1270 & 0
\end{pmatrix},$$

$$U[:, 6:9] = \begin{pmatrix}
-0.0157 & -0.1441 & 0 & 0 \\
0.2468 & 0.1570 & -0.6356 & 0.3099 \\
0.0066 & 0.0493 & 0 & 0 \\
-0.4842 & 0.8402 & 0 & 0 \\
0.2468 & 0.1570 & 0.6356 & -0.3099 \\
0.2287 & -0.0338 & -0.3099 & -0.6356 \\
0.0066 & 0.0493 & 0 & 0 \\
-0.7340 & -0.4657 & 0 & 0 \\
0.2287 & -0.0338 & 0.3099 & 0.6356
\end{pmatrix},$$

$$\Sigma = \begin{pmatrix}
1.5777 & 0 & 0 & 0 & 0 & 0 & 0 \\
0 & 1.2664 & 0 & 0 & 0 & 0 & 0 \\
0 & 0 & 1.1890 & 0 & 0 & 0 & 0 \\
0 & 0 & 0 & 0.7962 & 0 & 0 & 0 \\
0 & 0 & 0 & 0 & 0.7071 & 0 & 0 \\
0 & 0 & 0 & 0 & 0 & 0.5664 & 0 \\
0 & 0 & 0 & 0 & 0 & 0 & 0.1968 \\
0 & 0 & 0 & 0 & 0 & 0 & 0 \\
0 & 0 & 0 & 0 & 0 & 0 & 0
\end{pmatrix},$$

$$V[:, 1:5] = \begin{pmatrix}
0.1680 & -0.4184 & 0.6005 & -0.2256 & 0 \\
0.4471 & -0.2280 & -0.4631 & 0.2185 & 0 \\
0.2687 & -0.4226 & -0.5009 & -0.4900 & 0 \\
0.3954 & -0.3994 & 0.3929 & 0.1305 & 0 \\
0.4708 & 0.3028 & 0.0501 & 0.2609 & 0.7071 \\
0.3162 & 0.5015 & 0.1210 & -0.7128 & 0 \\
0.4708 & 0.3028 & 0.0501 & 0.2609 & -0.7071
\end{pmatrix},$$

$$V[:,6:7] \;=\; \begin{pmatrix} 0.5710 & -0.2432 \\ 0.4872 & 0.4986 \\ -0.2451 & -0.4451 \\ -0.6132 & 0.3697 \\ -0.0113 & -0.3405 \\ 0.0166 & 0.3542 \\ -0.0113 & -0.3405 \end{pmatrix}.$$

This matrix A (of rank $r_A = 7$) has 7 nonzero singular values, and the 2 trailing zero rows of the diagonal matrix Σ indicate that the first 7 columns of U (i.e., U_1, U_2, $\ldots$, U_7) determine a basis for the column space of A. From equation (4.7), we know that $\|A - A_6\|_F = \sigma_4 = 0.1968$ so that $\|A - A_6\|_F / \|A\|_F \approx 0.0744$ for $\|A\|_F = 2.6458$. Hence, producing a rank-6 approximation to the original matrix A reflects only a 7% relative change from A. Reductions to rank-5 and rank-3 would reflect relative changes of 23% and 46%, respectively. If changes of 46% are deemed too large (compared with the initial uncertainty in the original term-by-document matrix A), then modest rank reductions (say from 7 to 5) may be more appropriate for this small text collection.

To accept a 5-dimensional subspace and the corresponding matrix A_5 as the *best* representation of the semantic structure of a collection/database implies that any further rank reduction will not model the *true* content of the collection. The choice of rank that produces the *optimal* performance of LSI (for any database) remains an open question and is normally decided via empirical testing [6, 7]. For very large databases, the number of dimensions used may be between 100 and 300 [39]. The choice here is typically governed by *computational feasibility* as opposed to accuracy. Keep in mind that constructing the rank-k matrix A_k via SVD factors does produce the *best* approximation regardless of the choice of k.

4.2.1   Low-Rank Approximations

Recall from section 4.1 that relative changes of 32% and 76% were required to reduce the rank of the matrix A from Figure 3.1 to 5 and 4, respectively. Using the SVD, the relative changes required for reductions to ranks 5 and 4 are considerably less, i.e., 23% and 35%. As pointed out in [6], a visual comparison of the rank-reduced approximations to A can be misleading.

Notice the similarity of the rank-4 QR-based approximation ($\tilde{A}_4$) to the original term-by-document matrix A in Figure 4.2. The more accurate SVD-based approximation (A_4) is only mildly similar to A.

It is interesting to note that (by construction) the term-by-document matrix A will have nonnegative elements (weighted frequencies) (see section 3.1.1). From Figure 4.2, we see that both A_4 and $\tilde{A}_4$ have negative elements (which reflect various linear combinations of the elements of the matrix A). While this may seem problematic, the individual components of document vectors (columns of A_k) are not typically used to define semantic content. In other words, the *geometric* relationships between vectors (e.g., cosines) in the vector space are used to model concepts spanned by both terms and documents (see section 3.2.3). Notice, for example, that the first document vector (column) in the rank-4 matrix A_4 has (positive) components associated with terms that did not occur in the original document D1: $A_4(4,1) = 0.1968$ and $A_4(8,1) = 0.2151$. The vector representation of this document (D1: *Infant & Toddler First Aid*) now has components associated with the relevant terms T4: *Health* and T8: *Safety*. This ability to automatically associate related terms (without human intervention) is the hallmark of vector space modeling and motivation for low-rank approximation techniques.

4.2.2 Query Matching

As was done for the QR factorization in section 4.1, query matching can be formulated using the component matrices of the SVD. This particular formulation is the foundation of vector space IR models such as LSI (see chapter 3). Suppose we want to compare the query vector q with the columns of the reduced-rank matrix A_k (as opposed to the original $m \times n$ term-by-document matrix A). Suppose the vector e_j denotes the jth canonical vector of dimension n (i.e., the jth column of the $n \times n$ identity matrix I_n). Then, it follows that the vector $A_k e_j$ is simply the jth column of the rank-k matrix A_k. Similar to equation (4.5), the cosines of the angles between the query vector q and the n document vectors (or columns) of A_k can be represented by

$$\cos \theta_j = \frac{(A_k e_j)^T q}{\| A_k e_j \|_2 \, \| q \|_2} = \frac{(U_k \Sigma_k V_k^T e_j)^T q}{\| U_k \Sigma_k V_k^T e_j \|_2 \, \| q \|_2} \tag{4.8}$$

$$= \frac{e_j^T V_k \Sigma_k (U_k^T q)}{\| \Sigma_k V_k^T e_j \|_2 \| q \|_2} \text{ for } j = 1, 2, \ldots, n.$$

For the *scaled* document vector $s_j = \Sigma_k V_k^T e_j$, the formula in equation (4.8) can be simplified to

$$\cos \theta_j = \frac{s_j^T (U_k^T q)}{\| s_j \|_2 \| q \|_2}, \quad j = 1, 2, \ldots, n. \tag{4.9}$$

This implies that one need not explicitly form the rank-k matrix A_k from its SVD factors (U_k, Σ_k, V_k) and that the norms $\| s_j \|_2$ can be computed once, stored, and retrieved for all query processing.

The k elements of the vector s_j are the coordinates of the jth column of A_k in the basis defined by the columns of U_k. In addition, the k elements of the vector $U_k^T q$ are the coordinates in that basis of the projection $U_k U_k^T q$ of the query vector q into the column space of A_k [6]. An alternate formula for the cosine computation in equation (4.9) is

$$\cos \hat{\theta}_j = \frac{s_j^T (U_k^T q)}{\| s_j \|_2 \| U_k^T q \|_2}, \quad j = 1, 2, \ldots, n, \tag{4.10}$$

where the cost of computing the projected query vector $U_k^T q$ is usually minimal (i.e., q is typically sparse if the user supplies only a few search terms). For all document vectors (s_j), $\cos \hat{\theta}_j \geq \cos \theta_j$ so that a few more relevant documents may sometimes be retrieved if equation (4.10) rather than equation (4.9) is used.

4.2.3   Software

In order to compute the SVD of sparse term-by-document matrices, it is important to store and use only the nonzero elements of the matrix (see section 3.2.2). Numerical methods for computing the SVD of a sparse matrix include iterative methods such as Arnoldi in [38], Lanczos in [34, 47], subspace iteration [47, 48], and trace minimization [52]. All of these methods reference the sparse matrix A only through matrix–vector multiplication operations, and all can be implemented in terms of the sparse storage formats discussed in section 3.2.2. Chapter 9 provides several references and websites that provide both algorithms and software for computing the SVD of sparse matrices.

The original 9×7 term-by-document matrix A is

$$\begin{pmatrix} 0 & 0.5774 & 0 & 0.4472 & 0.7071 & 0 & 0.7071 \\ 0 & 0.5774 & 0.5774 & 0 & 0 & 0 & 0 \\ 0 & 0 & 0 & 0 & 0 & 0.7071 & 0.7071 \\ 0 & 0 & 0 & 0.4472 & 0 & 0 & 0 \\ 0 & 0.5774 & 0.5774 & 0 & 0 & 0 & 0 \\ 0.7071 & 0 & 0 & 0.4472 & 0 & 0 & 0 \\ 0 & 0 & 0 & 0 & 0.7071 & 0.7071 & 0 \\ 0 & 0 & 0.5774 & 0.4472 & 0 & 0 & 0 \\ 0.7071 & 0 & 0 & 0.4472 & 0 & 0 & 0 \end{pmatrix}.$$

The rank-4 approximation $(\tilde{A}_4)$ computed using the QR factorization is

$$\begin{pmatrix} 0 & 0.5774 & 0 & 0.4472 & 0.5983 & 0 & 0.5983 \\ 0 & 0.5774 & 0.5774 & 0 & 0.0544 & 0 & 0.0544 \\ 0 & 0 & 0 & 0 & 0 & 0 & 0 \\ 0 & 0 & 0 & 0.4472 & 0.2176 & 0 & 0.2176 \\ 0 & 0.5774 & 0.5774 & 0 & 0.0544 & 0 & 0.0544 \\ 0.7071 & 0 & 0 & 0.4472 & 0 & 0 & 0 \\ 0 & 0 & 0 & 0 & 0 & 0 & 0 \\ 0 & 0 & 0.5774 & 0.4472 & -0.1088 & 0 & -0.1088 \\ 0.7071 & 0 & 0 & 0.4472 & 0 & 0 & 0 \end{pmatrix},$$

and the rank-4 approximation (A_4) via the SVD is

$$\begin{pmatrix} -0.0018 & 0.5958 & -0.0148 & 0.4523 & 0.6974 & 0.0102 & 0.6974 \\ -0.0723 & 0.4938 & 0.6254 & 0.0743 & 0.0121 & -0.0133 & 0.0121 \\ 0.0002 & -0.0067 & 0.0052 & -0.0013 & 0.3569 & 0.7036 & 0.3569 \\ 0.1968 & 0.0512 & 0.0064 & 0.2179 & 0.0532 & -0.0540 & 0.0532 \\ -0.0723 & 0.4938 & 0.6254 & 0.0743 & 0.0121 & -0.0133 & 0.0121 \\ 0.6315 & -0.0598 & 0.0288 & 0.5291 & -0.0008 & 0.0002 & -0.0008 \\ 0.0002 & -0.0067 & 0.0052 & -0.0013 & 0.3569 & 0.7036 & 0.3569 \\ 0.2151 & 0.2483 & 0.4347 & 0.2262 & -0.0359 & 0.0394 & -0.0359 \\ 0.6315 & -0.0598 & 0.0288 & 0.5291 & -0.0008 & 0.0002 & -0.0008 \end{pmatrix}.$$

Figure 4.2: The term-by-document matrix A and its two rank-4 approximations $(A_4, \tilde{A}_4)$.

For relatively small term-by-document matrices, one may be able to ignore sparsity altogether and consider the matrix A as *dense*. The LA-PACK [1] Fortran library provides portable and robust routines for computing the SVD of dense matrices. Matlab [41] provides the dense SVD function [U,Sigma,V]=svd(A) if A is stored as a dense matrix, and [U,Sigma,V]= svd(full(A)) if A is stored as a sparse matrix. Matlab (version 5.1) also provides a function to compute a few of the largest singular values and corresponding singular vectors of a sparse matrix. If the k largest singular values and corresponding left and right singular vectors are required, the Matlab command [Uk,Sigmak,Vk] = svds(A,k) can be used. The sparse SVD function svds is based on the Arnoldi methods described in [37]. Less expensive factorizations such as QR (section 4.1) and the ULV decomposition [11] can be alternatives to the SVD, whether A is dense or sparse.

4.3   Semidiscrete Decomposition (SDD)

As discussed in [6], no genuine effort has been made to preserve sparsity in the reduced rank approximation of term-by-document matrices. Since the singular vector matrices are often dense, the storage requirements for U_k, Σ_k, and V_k can vastly exceed those of the original term-by-document matrix. The SDD [31] provides one means of reducing the storage requirements of IR models such as LSI. In SDD, only the three values $\{-1, 0, 1\}$ (represented by two bits each) are used to define the elements of U_k and V_k, and a sequence of integer programming problems are solved to produce the decomposition. These mixed integer programming problems with solutions (triplets) of the form $\{\sigma_k, u_k, v_k\}$ can be represented by

$$\min_{\substack{u \in \mathcal{S}^m \\ v \in \mathcal{S}^n \\ \sigma > 0}} F_k(\sigma, u, v) \equiv \|R_k - \sigma u v^T\|_F^2, \qquad (4.11)$$

where $\mathcal{S}^j$ denotes the j-dimensional subspace spanned by vectors whose components are in the set $\{-1, 0, 1\}$, $R_k = A - A_{k-1}$, and $A_0 \equiv 0$. As discussed in [31], the storage economization achieved by the SDD can result in almost a 100% reduction in memory required by the term (U_k) and document (V_k) vectors at the cost of a much larger (sometimes as much as 50%) rank (k) and subsequent computational time.

4.4   Updating Techniques

Unfortunately, once you have created an index using a matrix decomposition such as the SVD (or SDD) it will probably be obsolete in a matter of seconds. Dynamic collections (e.g., webpages) mandate the constant inclusion (or perhaps deletion) of new information. For vector space IR models based on the SVD (e.g., LSI), one approach to accommodate additions (new terms or documents) is to recompute the SVD of the new term-by-document matrix, but, for large collections, this process can be very costly in both time and space (i.e., memory). More tractable approaches such as folding-in, SVD-updating, and SDD-updating are well documented [6, 7, 31, 45, 56]. The procedure referred to as *folding-in* is fairly inexpensive computationally but typically produces an inexact representation of the updated collection. It is generally appropriate to *fold-in* only a few documents (or terms) at a time. Updating, while more expensive, preserves (or restores) the representation of the collection as if the SVD (or similar decomposition) had been recomputed. We will briefly review the folding-in procedure in light of our discussion on query matching in section 4.2.2 and provide further (comprehensive) reading material on updating the decompositions of sparse term-by-document matrices in chapter 9.

Folding a new document vector into the column space of an existing term-by-document is synonymous with finding the coordinates for that document in an existing basis (U_k). To fold a new $m \times 1$ document vector $\hat{p}$ into the (k-dimensional) column space of an $m \times n$ term-by-document matrix A means to *project* $\hat{p}$ onto that space [6]. If p represents the projection of $\hat{p}$, then it follows (see section 4.2.2) that

$$p = U_k U_k^T \hat{p}.$$

Hence, the coordinates (k of them) for p in the basis U_k are determined by the elements of $U_k^T \hat{p}$.

The new document is then *folded-in* by appending (as a new column) the k-dimensional vector $U_k^T \hat{p}$ to the existing $k \times n$ matrix $\Sigma_k V_k^T$, where n is the number of previously indexed documents. In the event that the matrix product $\Sigma_k V_k^T$ is not explicitly computed, we can simply append $\hat{p}^T U_k \Sigma_k^{-1}$ as a new row of V_k to form a new matrix $\hat{V}_k$. The product $\Sigma_k \hat{V}_k$ is the desired result, and we note that the matrix $\hat{V}_k$ is no longer orthonormal. In fact, the row space of the matrix $\hat{V}_k^T$ does not represent the row space of the new term-by-document matrix. In the event that the new document vector

$\hat{p}$ is *nearly* orthogonal to the columns of U_k, most information about that document is lost in the projection p. From section 3.2.3, we illustrate the folding-in of the new document

D8: Safety Guide for Child Proofing Your Home

into the rank-2 (SVD-based) approximation of the 9×7 term-by-document matrix A in Figure 3.1. For D8, we define

$$\hat{p} = q/\| q \|_2 \text{ for } q = (0 \quad 1 \quad 1 \quad 0 \quad 1 \quad 0 \quad 1 \quad 1 \quad 0)^T,$$

so that

$$\hat{p} = (0 \quad 0.4472 \quad 0.4472 \quad 0 \quad 0.4472 \quad 0 \quad 0.4472 \quad 0.4472 \quad 0)^T.$$

The two coordinates for the projected document (p) are defined by

$$p = U_2^T \hat{p} = (0.6439 \quad -0.0128)^T,$$

where the rank-2 SVD-based IR model is $A_2 = U_2 \Sigma_2 V_2^T$. Figure 4.3 illustrates the projection of D8 into the same 2-dimensional vector space depicted in Figure 3.3.

To fold in an $n \times 1$ term vector $\hat{w}$ whose elements specify the documents associated with a term, $\hat{w}$ is projected into the row space of A_k. If w represents the term projection of $\hat{w}$, then

$$w = V_k V_k^T \hat{w}.$$

The k coordinates $V_k^T \hat{w}$ of the projected vector w are then appended (as a new row) to the matrix U_k. In this case, we lose orthogonality in our k-dimensional basis for the column space of A (see [7, 45] for further details).

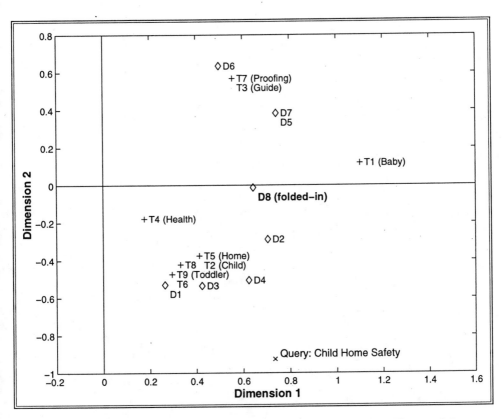

Figure 4.3: Folding D8 into the rank-2 LSI model from Figure 3.3.

Chapter 5

Query Management

Although a good portion of this book has been focused on the use of vector space IR models (e.g., LSI), it is important to continually remind ourselves and readers alike that there are several proven approaches used to build search engines. Other indexing models have certain strengths and limitations, and this becomes evident when one looks at query management or the process of translating the user's query into a form that the search engine can utilize. The relationship between the system and the query has a Catch-22 flavor. Selecting the most efficient search engine may depend on the type of query allowed which in turn depends on the indexing model used by a search engine.

In other words, certain types of search engine models handle certain types of queries better, but the user may have altogether some other type of query in mind. In this chapter, the focus will be on general guidelines for query management along with some discussion on the effects query types and search engines can have on each other.

5.1 Query Binding

Query binding is a general term describing the three-tier process of translating the user's need into a search engine query. The first level involves the user formulating the information need into a question or a list of terms using his or her own experiences and vocabulary and entering it into the search engine. On the next level, the search engine must translate the words (with possible spelling errors such as cran*beery*) into processing tokens (as discussed in section 2.4), and finally, the third level, the search engine must use the processing tokens to search the document database and retrieve the appropriate documents (see Figure 5.1).

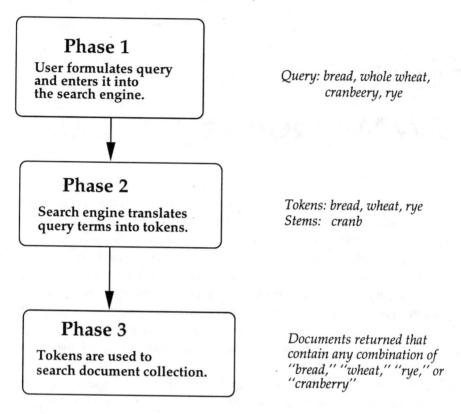

Figure 5.1: Three phases of query binding.

In practice, users (based on their work experience and skill level) may enter their query in a variety of different ways such as a Boolean statement, as a question (*natural language* query), or as a list of terms with proximity operators and contiguous word phrases, or they may use a thesaurus. Problems therefore arise when search engines cannot accept all types of queries [14]. A query that uses a Boolean operator such as *AND* or *OR* has to be processed differently than a natural language query. Building a search engine to accept a certain type of query ultimately forces the user to *learn* how to enter queries in that manner.

This is not to say that one type of query is better than another but rather to point out that there are several types of queries, and search engine design is extremely dependent on what type of queries the search engine will accept. Part of choosing the type of query is also dependent on anticipating who will be using the search engine. A beginning or inexperienced searcher such as a high school student will probably be more comfortable with a natural language query whereas an information professional who is familiar with advanced search features will be more comfortable with a system that can perform Boolean searches or searches using proximity parameters and contiguous word phrases.

5.2 Types of Queries

There are several types of queries. Each type should be evaluated with respect to how the user enters it and what he or she expects in return, its strengths and limitations, and compatibility with the search engine design. There are no hard and fast rules when discussing query/search engine compatibility. It is also interesting to note that the current trend for most operational systems is to apply a combination of queries with the hope that users will learn the best strategies to use for a given system [40]. Some systems are more *hybrid* than others and offer information seekers choices between exact (lexical) match Boolean and ranked approaches.

5.2.1 Boolean Queries

Boolean logic queries link words in the search using operators such as *AND*, *OR*, and *NOT*. Recall the document file in section 2.5 (the Bread Search limerick). If the searcher wanted the document that contains the words *meat OR wheat*, documents three, six, and seven would be pulled, but if the searcher wanted *meat AND wheat*, no documents would have been pulled. Why? Because none of the lines of the limerick contain both words *meat* and *wheat*. (To keep your Boolean operators straight remember the adage, "*OR* gets you more.") The *NOT* operator is more troublesome, because it forces the system to exclude members from the pulled or returned set of results. One of the weaknesses of Boolean queries, according to [32, p. 55], is that there doesn't appear to be a good way to gauge significance in a Boolean query. Either a term is present or it is absent. Current research [32] suggests that most users of information systems are not well trained in Boolean operators (unlike computer scientists and mathematicians). For systems based on a vector space IR model, Boolean operators are typically recognized as stop words, and hence are ignored.

5.2.2   Natural Language Queries (NLQs)

Natural language queries are queries which the user formulates as a question
or a statement. Using the Bread Search limerick from section 2.5, a viable
NLQ might be

> Which documents have information on manna?

or

> Find me information on the use of manna as a connotation for
> spirituality.

To process an NLQ, the search engine must extract all indexed terms to
initiate a search. Obviously, some words in the query will be eliminated by
the use of stop lists. This approach, according to Korfhage in [32], has the
distinct disadvantage in that "[C]omputers have difficulty extracting a term
with its syntactic, semantic and pragmatic knowledge intact."

In other words, once you extract a word from an NLQ, the *context* of
how that word is used becomes lost.

For example, in the NLQ query once the word *manna* is removed from
the query, then the system doesn't know whether the meaning is *manna
for bread* or *manna for spiritual sustenance*. This is not a major concern
perhaps in the case of *manna*, but when a word can have multiple and varied
meanings (i.e., when it is polysemous), then the effect can be much more
pronounced.

5.2.3   Thesaurus Queries

A *thesaurus query* is one where the user selects the term from a previous set
of terms predetermined by the IR system. The advantage of the thesaurus
query is that the first phase of query binding (see Figure 5.1) is automated
for the user. Also, thesauri can readily expand to related concepts. On
the other hand, the searcher is bound to the thesaurus term even if he or
she doesn't think the term is the *best* choice. Furthermore, there can be
problems with understanding the specific meaning of a word. For example,
in the Bread Search limerick, if the user is using the thesaurus to search
on the spiritual aspects of *manna*, the system may only consider manna as
bread described in the Old Testament of the Bible.

As mentioned by Kowalski in [33], thesauri tend to be generic to a language and can thus introduce/insert many search terms that were never

indexed or found in the collection. This is less of a problem in an IR system where the thesaurus has been especially built for the database; however, the costs to do so may be prohibitive.

5.2.4   Fuzzy Queries

It seems appropriate that the concept of *fuzzy searches* is somewhat *fuzzy* itself. Fuzzy reflects nonspecificity and can be viewed in two ways. As described in [33], fuzziness in a search system refers to the capability of handling the misspelling or variations of the same words. User queries are automatically stemmed and documents related to (or containing) the word stems are retrieved. Using the Bread Search document collection from section 2.5, a fuzzy query, which misspelled the word *cranberry* as *cranbeery* would still (it is hoped) return documents on cranberries, as the word stem (root) *cranb* would be the same for both words.

Fuzzy queries can also be viewed in terms of the sets of retrieved documents. It is similar to standard retrieval systems in which the retrieved documents are returned by order of relevancy. However, in the fuzzy retrieval system the threshold of relevancy is expanded to include additional documents, which may be of interest to the user (see [32]).

5.2.5   Term Searches

Perhaps the most prevalent type of query (especially on the WWW) is when a user provides a few words or phrases for the search. More experienced searchers might prefer using a *contiguous* word phrase such as *Atlanta Falcons* to find information about that football team. Otherwise, if the search engine does not interpret the query as a phrase, documents about the city of *Atlanta* or documents about *birds* would more than likely be erroneously retrieved. The use of contiguous phrases improves the chance of precision (see section 6.1.1) at the risk of losing a few documents that might have been selected. For example if an article read,

In Atlanta, the Falcons host their arch rivals...

it would not be retrieved (lower recall). In the next chapter on ranking and relevance feedback, the issues of precision and recall will be discussed further.

On the other hand, the experienced searcher may use a *proximity* operator, i.e., designate to the system that he or she wants a document that has the words *Atlanta* and *Falcons* with fewer than five words between them. (In most systems, users can choose five words, ten words, and so on).

As mentioned in section 2.5, in order for systems to support proximity operators and contiguous phrases, the inversion list not only must track which terms appear in which documents but also the position of those terms in the documents. This certainly requires more storage and processing capabilities.

One interesting dilemma for the user is deciding how many terms to supply. Users have a tendency to type in just two or three terms. This would be fine if the system was more match oriented or based on the Boolean *OR* operator. However, if the IR system is based on a vector space model, the user may have more success with more terms. In this context, more terms means more vectors to synthesize and construct the final query vector. However, the user may not realize that the query (in this case) could be more of a *pseudo*-document representation for conceptual matching as opposed to a long string of terms for lexical matching (see [6, 7, 18]).

5.2.6   Probabilistic Queries

Probabilistic queries refer to the manner in which the IR system retrieves documents according to relevancy. The premise for this type of search is that for a given document and a query it should be possible to compute a probability of relevancy for the document with respect to the query. In [32, p. 88], the difference between Boolean- and vector-based matching and probabilistic matching is summarized as follows:

> One concern about both Boolean- and vector-based matching is that they are based on *hard* criteria. In Boolean-based matching, either a document meets the logical conditions or it does not; in vector-based matching, a similarity threshold is set, and either a document falls above that threshold or it does not. The ability to define various similarity measures and thresholds in the vector case softens the impact of the threshold value to some extent. However, a number of researchers, believing that even this does not adequately represent the uncertainties that abound in text retrieval, focus their attention on models that include uncertainties more directly.

Marchionini in [40] acknowledges that in probabilistic IR modeling that there are degrees (or levels) of relevance that can be based on some estimated probability of document-query relevance. One advantage of probabilistic queries over fuzzy queries is that there is a well-established body of methods for computing probabilities from frequency data [32, p. 71]. For additional reading on probabilistic IR models (theory and empirical testing), we refer the reader to [15, 35].

Chapter 6

Ranking and Relevance Feedback

Even though at first it seems that all search engines have the same basic features, one aspect that separates fuller functional search engines from their lesser counterparts is the ability not only to list the search results in a meaningful way (ranking), but also to allow the user a chance of using those results to resubmit another, more on-target query. This latter feature, referred to as *relevance feedback*, has been quite effective in helping users find more relevant documents with less effort [50]. Harman [29] has done extensive work in judging how relevance feedback positively affects system performance.

Marchionini [40] points out another advantage of ranking and relevance feedback. As computer designers continue in their quest to make computers more receptive to users, ranking and relevance feedback can be considered as *highly interactive* information seeking. Along the same vein, Korfhage [32] refers to relevance feedback as a type of *dialogue* between user and system.

Some commercial systems limit this feature and elect to forego building it into the system since relevance feedback can create computational burdens that can slow a system down. Also, it requires constant updating of the system's knowledge of itself because in order for the search engine to know if a certain document is the best it must know what every document in the system contains. By using applied mathematics, producing a rank-ordered list of documents, and relevance feedback, search engine performance can be improved. Subsequently, vector space IR models such as LSI (see chapter 3) can be more readily adapted to providing relevance feedback. According to Marchionini,

The vector approach has a significant advantage over traditional indexing methods for end users, because the retrieved sets of documents can be ranked, thus eliminating the *no hits* result so common in exact-match systems. Experimental systems that provide ranked output have proved highly effective and commercial vendors have begun to offer ranked output features. Ranked output also provides a reasonable entry point for browsing. [40, p. 25]

6.1 Performance Evaluation

A variation of the sometimes pointless adage, "It doesn't matter whether you win or lose, it's how you play the game," is applicable in evaluating the performance of search engines. In one aspect, the ultimate evaluation of any search system is determined by whether the user is satisfied with results of the search. In short, has the information need been met in a timely manner? If so, the system and the user alike are declared the winner.

But this kind of simplistic thinking is, well, too simplistic. User satisfaction can certainly be measured in a variety of ways: *binary* (i.e., the results are acceptable or unacceptable), or *relative* (e.g., ranks a 5 on a scale of 1 to 10 with 10 being perfect). Also, other considerations need to be addressed when evaluating a system, such as whether the user didn't get the desired results because of a poor query in the first place. A misspelled word, the need for quotation marks to be added to a phrase, or lack of understanding of exactly what needs to be asked can also affect a search engine's performance.

Korfhage [32] also points out that users won't tolerate more than three or four attempts of feeding back information to the system, nor does the user like seeing the same set of documents appearing over and over again. This brings out an interesting dilemma for the system designer. For example, after receiving a list of documents, suppose the user marks a certain document as *retrieve more like this one*. In the new return list of results, should that document be automatically at the top of the next list even though conceivably it could be ranked lower than the new set of documents? Then again, if the document isn't at the top of the list the user may wonder why.

Such problems, however, have not prevented researchers from developing some standards for at least discussing and comparing the results of a search whether it be searching on the web, an on-line database, or a CD-ROM.

6.1.1 Precision

Unlike information science professionals who are drilled early in their academic careers by the standard definitions of precision and recall, system designers may not be as familiar with the terms and the implications of their use. This is understandable because it's one thing to design and build a system; it's another thing to evaluate it. *Precision* and *recall* are two of the standard definitions used in search system evaluation, and because they are so closely related they are usually discussed in tandem.

The precision or precision ratio P of a search method is defined as

$$P = \frac{D_r}{D_t},\tag{6.1}$$

where D_r is the number of relevant documents retrieved and D_t is the total number of documents retrieved. It's important to keep in mind that *relevance*, or the appropriateness of a document to a user's needs, is still a judgment call. Two searchers can be looking for the same information in the same topic and when the results are listed, *Article* 4 may strike *searcher A* as useful or relevant, whereas *searcher B* may find *Article* 4 to be irrelevant and not useful. Relevance is subjective and dependent on who is keeping score. Sometimes it is the end user. Sometimes it's an intermediator (an information broker, for example) who is making that determination for someone else, and sometimes it's a disinterested third party system evaluator who is simply evaluating search engines (and has no vested interest in the information itself).

6.1.2 Recall

The recall or recall ratio R of a search method is defined as

$$R = \frac{D_r}{N_r},\tag{6.2}$$

where D_r is the same numerator from equation (6.1) and N_r is the total number of relevant documents in the collection. Recall ratios are somewhat

difficult to obtain, as the *total* number of relevant documents is always an
unknown. In other words, how does one compute the recall ratio if you're
not really sure how many relevant documents are in the collection in the
first place? However, that doesn't lessen the usefulness of the recall ratio.

In one sense, precision and recall are on a continuum. If a searcher wants
only the precise documents that fits his or her exact needs, then the query
will require very specific terms. However, there is the danger or trade-off that
if the search is extremely precise, many relevant documents will be missed.
That explains the integral role of recall. The search must be broadened
so that a significant number of relevant documents will be included in the
results of the search. Again there is a another trade-off: if the recall is
increased, the user more than likely will have to wade through what is known
as *false drops, garbage,* and *noise* to pick out the relevant documents. What
a user generally needs is a complete subset of relevant documents that does
not require a substantial *weeding out* of all the irrelevant material.

6.1.3   Average Precision

Average precision is a common measure used in the IR community to assess
retrieval performance [28]. Let r_i denote the number of relevant documents
up to and including position i in the (ordered) returned list of documents.
The recall at the ith document in the list (the first document is assumed
to be most relevant) is the proportion of relevant documents *seen* thus far;
i.e., $R_i = r_i/r_n$. The precision at the ith document, P_i, is defined to be
the proportion of documents up to including position i that are relevant to
the given query. Hence, $P_i = r_i/i$. The so-called pseudoprecision at a recall
level x is then defined by

$$\tilde{P}(x) = \max P_i, \text{where } x \leq \frac{r_i}{r_n}, \text{ and } i = 1, 2, \ldots, n. \qquad (6.3)$$

Using (6.3), the n-point (interpolated) *average precision* for a query is given
by

$$P_{av} = \frac{1}{n} \sum_{i=0}^{n-1} \tilde{P}\left(\frac{i}{n-1}\right).$$

As it is common to observe retrieval at recall levels $k/10$, for $k = 0, 1, \ldots, 10$,
an $n = 11$ point average precision (P_{av}) is typically used to measure the
performance of an IR system for each query. If a single measure is desired

for multiple queries, the mean (or median) P_{av} across all queries can be recorded.

6.1.4 Genetic Algorithms

Genetic algorithms can be used in conjunction with relevance feedback calculations to monitor several relevance feedback scenarios simultaneously before choosing the best one. The premise with this *combined* approach is that in the normal relevance feedback situation the user makes a decision early on, and many documents are quickly discarded. However, some of those discards could be relevant or lead to relevant documents, so in a system using a genetic algorithm those discarded documents are used to create *alternate paths* to run in parallel to the main set of queries. Later on, the system will determine whether the alternative sets of documents merit further consideration. Korfhage has done extensive work in genetic algorithms and has dedicated several sections of his book [32] to this topic.

6.2 Relevance Feedback

An *ideal* IR system would achieve high precision for all levels of recall. That is, it would identify all relevant documents without returning any irrelevant ones. Unfortunately, due to problems such as polysemy and synonymy (see chapter 3), a list of documents retrieved for a given query is hardly perfect, and the user has to discern which items to keep/read and which ones to discard.

Salton in [50] and Harman in [29] have demonstrated that precision can be improved using *relevance feedback*, i.e., specifying which documents from a returned set are most relevant so that those documents can be used to refine/clarify the original query. Relevance feedback can be implemented using the column space of the term-by-document matrix A (see section 3.1.2). More specifically, the query can be replaced or modified by the vector sum of the most relevant (user judged) documents returned in order to focus the search *closer* to those document vectors.

We will assume that the original query vector (q) lies within the column space of the matrix A_k, the low-rank approximation to the term-by-document matrix A (see sections 3.2.3 and 4.2). Otherwise, the query vector is replaced by its projection onto that column space ($U_k U_k^T q$ for SVD-encoded collections). For convenience, both query and document vectors can be represented using the same basis for the rank-k approximation to the column space of A [6]. For SVD-based approximations (section 4.2), this basis is determined by the columns of U_k, where $A_k = U_k \Sigma_k V_k^T$. The coordinates of the query vector are simply the elements of the vector $U_k^T q$, and the coordinates of the jth column $U_k \Sigma_k V_k^T e_j$ of the matrix A_k are the elements of the vector $\Sigma_k V_k^T e_j$.

If a user considered, say, the tenth document (a_{10}) of a collection to be most relevant for his or her search, then the new query q_{new} (based on relevance feedback) can be represented as

$$
\begin{aligned}
q_{new} &= U_k U_k^T q + a_{10} \\
&= U_k U_k^T q + U_k \Sigma_k V_k^T e_{10} \\
&= U_k (U_k^T q + \Sigma_k V_k^T e_{10}).
\end{aligned}
\tag{6.4}
$$

If the user judges a larger set of documents to be relevant, the new/modified query can be written as

$$
q_{new} = q + \sum_{j=1}^{d} w_j a_j = U_k (U_k^T q + \Sigma_k V_k^T w),
\tag{6.5}
$$

where the vector element w_j is 1 if a_j is relevant and 0 otherwise. If the user (or the IR system) desires to replace the original query by a sum of document vectors, the vector q in equation (6.5) is simply set to 0.

Note that the vector $U_k^T q$ in equations (6.4) and (6.5) can be formed in the original cosine computation with the original query vector q. Hence, the new query (q_{new}) can be determined efficiently via the sum of k-dimensional vectors and then compared with all documents vectors. Following equation (4.9), we can define the vector $s_j = \Sigma_k V_k^T e_j$ so that appropriate cosine formula becomes

$$
\cos \theta_j = \frac{s_j^T (U_k^T q_{new})}{\| s_j \|_2 \| q_{new} \|_2}
$$

for $j = 1, 2, \ldots, n$. As was shown in [20], experiments have shown that replacing a query with a combination of a few of the most relevant documents to the query returned can significantly improve the precision of LSI for some collections.

Chapter 7

User Interface Considerations

The concept of user interface, or creating the tool that people use to interact with the search engine, is an important part in the burgeoning field of *human–computer interaction* (or HCI). The significance of the user interface cannot be emphasized enough, because often the user will judge the performance of the search engine not on the final results of the search as much as what perceived *hoops* the user jumped through to get those results. For example, if it is difficult to type a search term in the small fill-in box, or there's uncertainty about how the engine will handle the search terms or relevant results seem questionable, then there is a possibility of user dissatisfaction. The importance of the interface between the user and the search engine cannot be overestimated.

It is certainly beyond the scope of this book to discuss all the related issues concerning user interfaces. Marchionini in [40] gives a broad view of the user-computer relationship. He focuses on *personal information infrastructure*, the mental model of the user and its composition of previous learning history, goals, level of expertise, expectations, etc. Shneiderman in [54] provides an excellent review of the *broad* field of HCI along with a general framework for the design of interfaces. According to Shneiderman, "[W]hen an interactive system is well-designed, the interface almost disappears, enabling users to concentrate on their work exploration, or pleasure"[54, p. 10].

7.1 Guidelines

In general terms, interface designers are encouraged to develop guidelines and goals of what the interface should do, i.e., what tasks/subtasks must be carried out for the users. An interface for a general user with limited search skills may need to be much different from that for a skilled searcher. The skilled searcher prefers to understand how the search operates in order to narrow or broaden it and to ensure its thoroughness, whereas the novice merely wants *an answer*. Marchionini believes systems should not only accommodate analytical, structured searchers, but he also sees the importance of more unstructured browsing (one of the early *advantages* of the Web), which can be considered a valuable type of learning method for many potential users. Moreover, the distinction between planned searching and simple browsing is often blurred. Therefore, says Marchionini, "A grand challenge for interface designers is to create new features that take advantage of the unique characteristics of each medium" [40, p. 45]. Following [54], some of the general features of an interface and how their integration might be accomplished are discussed below.

7.1.1   Interface Tools

Interface building tools are in a continual state of evolution. Although computer programs written in languages such as C and C++ are often the standard, more software tools are being developed that can shorten the time for initial layouts and subsequent revisions. Several *visual* programming tools, in particular, are now available to aid interface designers. Shneiderman [54, p. 166] has suggestions for popular commercial software products that can aid in overall interface design.

Although having the right tools for building the interface between the search engine and user is important, equally critical is deciding what features the interface needs to have.

7.1.2   Progress Indication

The *progress indicator* or some temporary onscreen notification is a necessary feature. In other words, while the search is being conducted the user needs to know what the system is doing with a *please wait* or *countdown of the seconds*, before the results arrive. Ironically, while it's important that results come quickly (e.g., within two to five seconds) there is a problem for the user if they come back *too fast*, i.e., instantaneously. This is easy to understand. For example, if the user types a search term, presses *Return*, and the engine comes back within a half second with *No results found*, a degree of skepticism emerges whether the engine ever searched or accessed anything (see shoe salesperson analogy below). Moreover, if the wait was only a second or two longer, the user would probably be satisfied that at least the search engine was executing properly.

Searching for Shoes

If you've ever worked as a shoe salesperson or waited tables for any length of time, you soon learn a valuable *trick* to keep the customer happy or at least avoid being hassled.

When a customer asks for a specific style and size of shoe or specific food item such as chocolate cheesecake that you know you're out of, the wise employee (rather than risk sounding brusque with a curt "We're out of that") will go the back room or kitchen and pretend to look anyway. Upon returning with the bad news, the customer is usually more understanding. After all, you gave it your best *search*.

The same is true with search engines. Come back instantaneously with *No results found* and you have a potentially dissatisfied user. Wait a few seconds and return the same message and the user knows at least the search engine *tried*.

7.1.3 No Penalties for Error

If the user perceives that any command can be *undone* or cannot *hurt the system*, they are more likely to be comfortable with the system. Along the same lines Shneiderman [54] also encourages designers if they use error messages to use ones that are not *harsh*. Instead of announcing *Syntax Error*, perhaps a less *threatening* suggestion would be *Please add a pair of closing quotation marks to your query*.

7.1.4 Form Filling

Although completing a form seems like a simple, logical, and easy task for the user to learn, designers should take steps to ensure that the form is appealing, easy to navigate, and void of any unnecessary text. This is especially important considering the screen limitations on some workstations and laptop computers. Although search engine interfaces don't require a lot of screen space to enter a query, some consideration should be given to the manner in which the user sends/invokes the query. For example, do you hit *Enter*, or press the *Search* icon? Can the user backspace or delete a misspelled word easily? Interestingly enough, if one looks at current industrial search engines (e.g., AltaVista, Excite, Infoseek, Yahoo!), the line/region where the query is typed is usually very small. Only a few characters can be typed in and it is difficult to see what was entered at the beginning of the query. Understandably, these commercial search engines have made a trade-off between advertising revenues and user friendliness, but it remains to be seen whether this short-term financial gain will evaporate in the long run. Frustration with a *busy* interface could lead to user avoidance altogether.

7.1.5   Display Considerations

While there are numerous layouts and designs that will attract user attention, it is advisable [54] that designers try to *control* themselves. For example, avoid more than two levels of intensity and use only a maximum of three fonts. Four standard colors are reasonable, and blinking text should be kept to a minimum.

7.1.6   No Anthropomorphic References

Experts in interface design such as Shneiderman tend to oppose the use of *anthropomorphism* or assigning human traits to inanimate objects. Current literature indicates that while anthropomorphism may seem *cute* at first, it soon becomes irritating and a constant reminder that the user is working with a computer, which defeats the purpose of anthropomorphism. According to Shneiderman [54], perhaps the *best* interface is the one that becomes the *most transparent*. Or to coin a phrase used in the sport of baseball, "The best umpires are the ones we don't notice."

Interaction Styles

Citing [53], Marchionini lists four types of interaction styles that are prevalent in search systems. Each type has certain characteristics that merit consideration:

Commands—the user types in specific commands that are specific to a particular system. Command systems are more difficult to learn, but search experts in general prefer them. A good example is DIALOG.

Menu—the user is presented a list of commands and selects which command he or she wants executed. The menu style interface allows the novice user the benefit of being able to work on the system without knowing the command language. However, users are limited in what actions they can take.

Form Fill-in—is a cross between menu and command styles of interaction.

Direct Manipulation—the user moves the mouse to enter a command and watches the results in real time. Direct manipulation systems are certainly one of the key components in the design of highly interactive yet *transparent* information-seeking systems [40].

7.1.7 Test and Retest

The interface should be tested and retested by individuals who were not part of the original design team. A related problem discussed by Marchionini in [40, p. 75] is the so-called user-centered design paradox: "We cannot discover how users can best work with systems until the systems are built, yet we should build systems based on knowledge of users and how they work."

Marchionini sees the advantage of continually studying and testing user interfaces. He also doesn't discount the importance of looking at patterns and strategies of novice users. In a way, they are the more natural users, and therefore, the interfaces should be adapted to fit their predilections instead of vice versa.

7.2   Interfaces for Search Engines

Both Marchionini and Shneiderman [54] reinforce the basic premise that the interface must be designed to meet a user's needs. This creates problems in search engine interfaces, because the range of expertise in search engine users varies from beginner to expert. Further complicating the interface design issues is that a user's information needs range from specific and related fact finding to browsing and exploration (to browse and explore means the search engine must have the ability to do concept searching, which is one of the strengths of vector space modeling). An interface must be able to handle both types of searches. This is no small task considering the limited space of a computer screen. Ideally, systems should initially be able to accommodate first time users, who as they gain experience with the system may require a much wider range of search tools for composing, saving, and revising queries.

7.2.1   User's Control

In order to improve the user interface, Shneiderman suggests that the system should allow the user to select different collections (if applicable), different fields (if applicable), and relax search constraints, such as case sensitivity, stemming, and partial matches. Also, the system should be able to recognize specific names and, if necessary, concepts. When the results are returned from the search, the list of items/documents should be able to be manipulated in terms of the number of results and the ability to change the order, whether it be alphabetically, chronologically, or in terms of relevancy. Moreover, according to Marchionini, how the results are displayed — whether it be chronologically or by query-term frequency or whether the terms are highlighted in the text — impact the user's decision on how to approach the next step in retrieving information.

7.2.2   Informing the User

Another facet of displaying the results is telling the user *how* the search engine processed the query [54]. Did the system interpret the user's query as the user intended? If so, was the user informed of this? For example, users may use wildcards, quotations to mark phrases, or Boolean operators (see chapter 5) such as *Or* and *And* in their query. However, since there is a lack of standardization in search engines, each search engine may handle the query differently. It is important to indicate how the query was processed without confusing the user.

7.3   In Practice

There are experts who have made careers out of studying, creating, and analyzing user interfaces. However, oftentimes search engine builders are asked or forced to build the user interface without all the experience they would probably like to have. But we all learn by doing (or at least imitating), and the computer science course, which was part of the impetus for writing this book, required five groups of students (i.e., mock Internet service provider (ISP) companies),

1. Search Busters,

2. Dig-In Products,

3. Information Pathways,

4. Safari Creations, and

5. Tachyon,

to build interfaces for their respective search engines (see chapter 8). It's interesting to note that the interfaces[4] developed by the students (who received little help with interface design) had many of the features described above.

The group *Information Pathways* informed the user which key words were searched along with the query results. They also used friendly faces and scowls to indicate which results were more relevant—borderline anthropomorphism? All the user interfaces had some way (such as a *pull-down* menu) for the user to indicate if they wanted the user's search to be more literal or conceptual (i.e., a dimension control for an implicit vector space IR model). With the *Dig-In Products* interface, the user received a flashing *processing status* message known in HCI vernacular as the *progress indicator* to indicate that the search engine was working on a query. All the class

---

[4]The CS460/594 (fall semester, 1997) course projects are available on the World Wide Web at http://www.cs.utk.edu/~cs460.d&im/cgi-bin/group.

project interfaces were simple and natural to operate. As alluded to above, these students were not *coached* on how to design a user interface. It was left entirely up to the groups. However, the interfaces were generally effective. This may indicate that if designers can briefly put themselves in their user's shoes, or include features *they like as a user*, that an effective intuitive user interface can begin to be designed. Further reading on interfaces for effective text searching is provided in [55].

Once all best efforts of interface building and evaluation have been exhausted, designers should take refuge in the philosophical words of encouragement from Marchionini who in [40] reminds interface designers that maximum efficiency isn't always agreeable to human nature. Although HCI's ultimate goal is to optimize performance, sometimes optimization may make tasks somewhat boring and impede performance. Marchionini maintains it is human nature to seek variety, even at the expense of doing things at maximum efficiency. In other words, sometimes the best way to do something is the way the user prefers, not the designer.

Chapter 8

A Course Project

In addition to the standard fare of lectures and exams, the central theme of a recent (fall semester 1997) senior undergraduate course, *Data and Information Management*, was the course project of building a search engine for a collection of webpages associated with federal statistics. With the help of Cathyrn Dippo at the Bureau of Labor Statistics, four sets of HTML pages (ASCII files) were obtained from agencies such as the Bureau of Labor Statistics, the Division of Science Resource Studies at the National Science Foundation, and the U.S. Census Bureau.

The intent was to give students in computer science and information sciences (at the University of Tennessee) a taste of the real-world experience of product development while utilizing material covered in lectures. However, there were two inherent problems with managing such a project within the framework of a computer science course.

1. If there were a large number of students (over 20), it would be difficult to give all of them enough hands-on time with the project.

2. There was a tight time frame; building a search engine is more than a weekend project.

Within a 16-week semester, the students would not have the luxury of attending all the lectures and studying all the class materials before starting to build their search engines. They would have to start from the first day of class, and students, since they are human like the rest of us, tend to procrastinate.

8.1   Project Approach

For this course project, Dr. Michael W. Berry adopted a modular approach
and divided the workload into five basic modules:

1. document preparation,

2. sparse term-by-document matrix construction,

3. implementation of the SDD algorithm (see section 4.3),

4. query formulation and document ranking, and

5. user interface construction.

Based on both experiences and backgrounds, responsibilities were assigned to a few selected individuals (jokingly referred to as the *Islanders*, or
islands unto themselves) to work on the document preparation, matrix construction, SDD algorithm, and query and ranking modules. The remaining
20 students were divided into five groups who were to construct independent search engines (including the user interface) using the five modules
listed above. The Islanders, who were the primary *software engineers* of the
project, volunteered for their duties. This is not an unusual phenomenon, as
some computer science students prefer to work alone on software projects.
However, software module development for this project did not relieve them
of the responsibility of interacting with other students in the class. In fact,
each Islander or software engineer (SE) and ISP group was required to submit a progress report, and on several occasions class time was used as a
forum for all the SEs (module designers) to field questions from the ISPs.

The small groups of students were instructed to design and implement
a product that might be marketed by a competitive ISP. Each mock ISP
in the class was given the same data/information to process and index for
query matching. The students were told they would be expected to present
their search engines to the class and a small panel of local experts for evaluation. Just knowing that their software would be evaluated by persons other
than the instructor and fellow students seemed to motivate the ISPs in a
somewhat competitive way.

In dividing the five groups, some consideration was made to make sure
that each group had one or two experienced programmers. Also, to ensure
student involvement, those with limited programming experience (e.g., information science students with some but not immediate expertise) were

encouraged to become especially involved with the designing and testing of the graphical user interface (GUI) and writing the help documentation. In some ways, this type of division of labor is not unlike what one might expect in a corporate product development experience, where individuals who bring various skills to a project must work together.

8.2 Time Frame

Having the students search in the same corpus of documents not only made for easier comparisons between search engines (similar to TREC), but it also meant that only one set of documents had to be prepared and *validated* for searching. Students were freed from the task of parsing the collection of webpages. An index file containing 5838 terms or key words was created from the 619 webpages on federal statistics. Stemming was not employed and a stop list of just under 500 ordinary words was used to discard unimportant terminology. The list of terms was sorted alphabetically and both local and global term weighting was applied to term frequencies. Parsing the document collection in advance was an attempt to deliberately shorten the development cycle for the ISPs. Having a graduate teaching assistant coordinate the distribution and sharing of modules created by the software engineers (Islanders) for use by the ISPs helped keep the process on track as well.

8.3 File Preparation

The original webpages were validated to ensure that each document had an opening <HTML> and closing <\HTML> tag. One SE volunteered to further *clean* the files. It was agreed that the internal anchors for local links inside these HTML files (HREF) were disabled; however, these internal anchors were left in the document because they could contain important search words. Links to other HTML documents outside the collection were left intact, so users could pull in those documents if they were still available. The SE volunteer used the Unix stream editor (sed) to identify and comment out all external image references preceded by the HTML tag <IMG>. The documents were then rearranged within the document file according to title, alphabetically. Documents were delineated by inserting a blank line before the title of each document (no blank lines were allowed within a document, of course). The SE inserted the blank lines using a small program written in C. In this way, the position (or line count) of each document could be calculated for future document retrieval by a search engine.

A very annoying problem with validating the HTML pages in the 619-webpage collection was an obvious lack of consistency with the tags. Tags specifying hypertext links had various forms: `<A hReF>`, `<A HREF>`, and `<a hreF>`. Consequently, the student was forced to make several passes through the file to identify all the possible forms. Although not used for this project, using a Unix shell or Perl script to identify all possible variations of HTML would be useful for HTML validation. Certainly, the document file preparation was critical in the project timetable, and every effort was made by the SE performing that task to finish early in the semester. The term-by-document matrix construction task could not begin until all webpages were validated.

8.4   Different Term-by-Document Matrices

After the webpages were validated and the terms/key words were extracted, another student (SE) wrote a C++ program to build a term-by-document matrix for each of the five selected weighting schemes. As described in section 3.2.1 and [31], the five weighting schemes (along with document normalization) used are listed in Table 8.1. By having each ISP use a different matrix encoding for the LSI model described in chapter 3, comparisons in term weighting could be implicitly made. The SE assigned with the matrix construction task created five different term-by-document matrices, which were then factored using the SDD described in section 4.3.

8.5   Query Processing

As mentioned earlier in chapter 3, the user's query can be represented as a vector (perhaps as a new column for the term-by-document matrix). The SE assigned to the query processing module used the term list generated from the original document file to essentially transform the user's query from typed words to an encoded vector. Key words from the user's query were compared with term list (implemented as a B-tree or similar data structure for fast searching). Key words or terms are encoded into the query vector using the global weights (g_i's) used to define the original term-

Table 8.1: Assigned term weights

Group No.	Local Weighting	Global Weighting
1	Logarithm	Inverse document frequency
2	Logarithm	Probabilistic inverse
3	Term frequency	Inverse document frequency
4	Term Frequency	Probabilistic inverse
5	Binary	Inverse document frequency

by-document matrix. Hence, the term list data structure should include the corresponding global weights so that query terms can be (globally) weighted to be consistent with the construction of a term-by-document matrix (prior to any factorization). Since most queries will involve multiple key words, the query vector may require normalization (similar to the columns of the term-by-document matrix). For example, the SE responsible for writing the query processing module may provide the option to normalize the query vector (i.e., divide all vector components) by the number of terms used to define the final query.

8.5.1 Ranking

The SE designed this module so that a precomputed factorization of the term-by-document matrix using the SDD (provided by another SE) could be used to compute similarities with the encoded query and all term and document vectors. Cosines were used for this project, but any other vector-based similarity measure could certainly be used. It was expected that the ISP groups would process the ranked items (terms, documents, or both) returned by the query processing module in such a way that the *best* match (highest similarity) would be presented first followed by items of reduced (but ordered) similarity.

It is important to realize that the ISPs did not require access to the final query processing module to develop their early prototypes. On the contrary, as a time saving measure, they were encouraged to generate simple random rankings of documents and make sure their interface could display any selected item in the *fabricated* ranking.

8.5.2   Relevance Feedback

Incorporating the relevance feedback module can be tricky since there are certain trade-offs that need to be resolved concerning the user interface and the way relevance feedback is actually implemented. Each mock ISP had to resolve questions of the following nature:

1. Should a new query replace the original query completely or just modify it?

2. Can previous terms/key words be deselected in the next query?

3. Should only the current list of returned documents be searched with a relevance feedback query, as opposed to searching the original document collection?

Unfortunately, most of the ISP groups did not implement relevance feedback in their search engines. This search feature was *not* a project requirement, and the students apparently did not have enough time to implement it correctly.

8.6   The User Interface

In a way, the ISP groups had a double-edged task. Not only were they required to program and interface with the module programmers (SEs), but they had to anticipate how a user would use the GUI they designed. They were also faced with technical concerns of deciding which languages to use (C++, CGI, Java) and what kind of HTML forms would be provided to the user. For the latter consideration, the five ISP groups had to determine how they would display the query results. At first one might think that the hyperlinked title would be enough, or the first few lines of an abstract, but the students noticed that some of the abstracts were not well written or incomplete. Also there were screen size limitations (see Figure 8.1) so that if the user interface displayed the results and parts of the full abstract, the user would require a lot of scrolling just to read the first 10 items from the returned list of documents. Another important consideration is how many documents/items for display should be retrieved from the collection at a time. The students quickly learned that excessive requests to the *http* server (machine) during peak times could be disastrous.

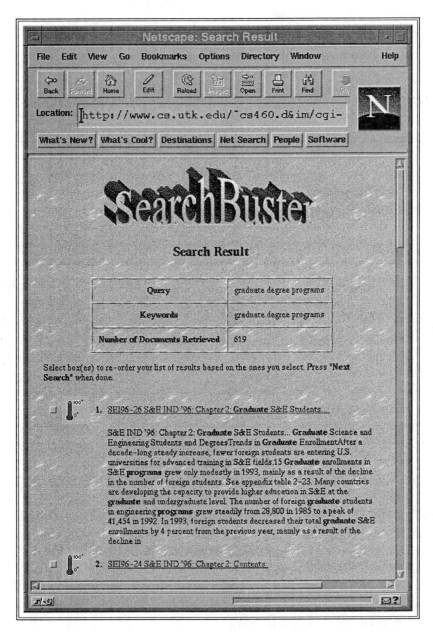

Figure 8.1: Sample search engine webpage that demonstrates how much screen space is occupied if full abstracts are displayed in the (ranked) return list of documents. Notice that the thermometer icon doesn't really help discern the similarities of the documents with the query *graduate degree programs*.

Other considerations for the user interface included whether to allow the user to change the dimensionality (k) of the vector space (or LSI) representation (see Figure 8.2). The higher the value of k, the more conceptual the search (see section 3), but this feature has to be explained to the user without confusion. Some of the ISPs chose to use the maximum (default) number of dimensions computed and prohibit any user modification.

The ISPs were also responsible for handling the ranking values (or scores). Some groups created icons (see Figure 8.3) to indicate which results had the higher rankings. Problems arose concerning whether to *bump up* a matching value of 60% similarity to 100% similarity when the respective result was the best match to the user's query. Another problem with similarity values relates to user interpretations. The statistical results between a 80% similarity compared with a 75% similarity is actually quite minimal, but the larger value will commonly be interpreted as the *best match*.

The ISP groups differed in how they elected to handle the user's typed-in query. Most groups displayed the query key words along with the results. However, at least one group (see Figure 8.4) displayed the key words but showed the user which words the system *did not accept*, and thus identified words from the stop list. Another difference among the interfaces involved how the user would add more words to the current query and resubmit the modified query. Some groups designed the HTML form to automatically clear, which could be a convenience or nuisance depending on how happy one was with the original key words. Another issue is whether the user has been informed that a modified query is essentially a new search entirely or just a slight revision of the current query.

By nature GUIs must look appealing, a matter subject to every user's taste and opinion. What is intuitive to one user is confusing to another. Each ISP group was forced to make decisions regarding these issues. Some of the groups had very little help text on the screen and kept their approach very basic. Others were more elaborate with interfaces that could potentially confuse the user in order to pack more information on one screen.

8.7   Project Evaluation

On the final day of class, the five ISPs presented their projects to fellow classmates and a panel of three judges. Evaluating a search engine doesn't necessarily require expertise in IR (although our group of judges was quite

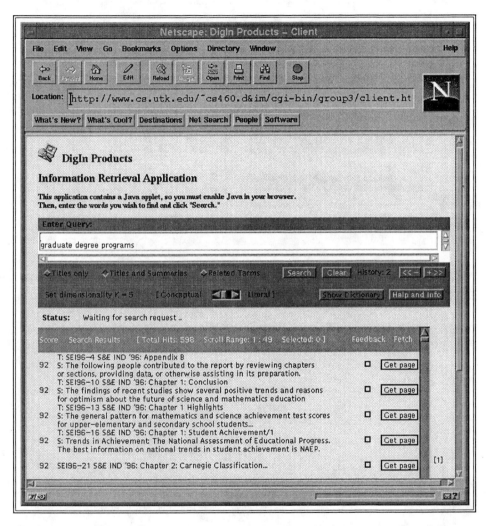

Figure 8.2: Sample search engine webpage that demonstrates the ability to change the dimensionality (k) of the vector space (LSI) model. The larger the value of k, the more *literal* the interpretation of the query.

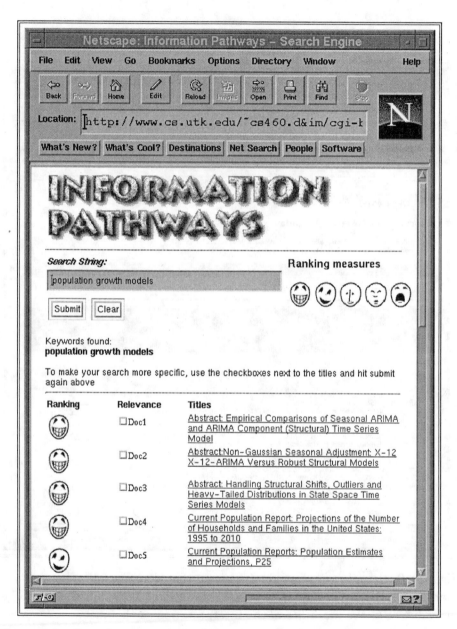

Figure 8.3: Sample return list of documents for the ISP group that used *face* icons to reflect similarities. The query was *population growth models*.

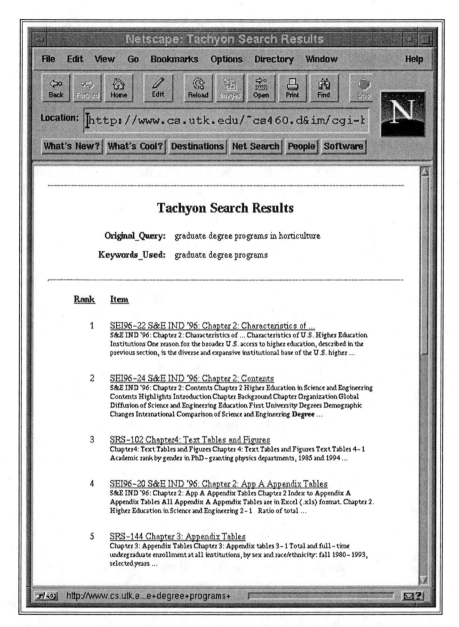

Figure 8.4: Sample return list of documents for the ISP group that indicated the actual key words found in the user's query. The query in this example was *graduate degree programs in horticulture*, and the first three words are the only key words.

knowledgeable). A reference librarian, a corporate researcher, or a person who has used any of the commercial net search engines on a regular basis would make a good evaluator. Designing a product for a potentially wide range of users should be evaluated from many different perspectives. The panel of judges for the project described for the fall 1997 course was asked to examine the five ISP search engines for level of effort, creativity, query-matching effectiveness, and in-class presentation.

The final evaluation was not intended to be another vehicle to assign a grade, rather it was presented as an opportunity for the students to demonstrate to their peers and to interested potential users what they could do. Keep in mind, the ISPs were instructed to be discrete in revealing their interface designs and approaches as much as possible during development. In a real-world environment it is imperative that designers understand discretion to avoid tipping off their competitors.

A few days before the final class, each ISP was required to make their search engine available for testing and evaluation by another group. This assignment served a twofold purpose. Not only would each ISP receive a constructive critique from an *understanding* peer, but students would also get some experience systematically evaluating other systems, which can be useful later on in the workplace when a programmer or information technology specialist may be asked to analyze a software program or system.

During the day of the presentation, students and judges heard each of the presentations followed by their respective critiquing group's evaluation. The evaluations were generally constructive and positive in nature. By this point in the semester, everyone knew that to insensitively criticize another's project was akin to attacking the designer's mother.

The groups varied their approach to the presentation. Sometimes the entire group stood up and *winged it* (which was okay because the students were instructed not to worry too much about their presentations) or in the case of one group, the most eloquent member of the group did the entire presentation while the programmer did the demonstration and answered any technical questions from the class, the evaluating group, and the judges. After the panel was sequestered to determine the final ranking of the search engines, students waited and discussed finals or future employment plans while eating doughnuts, which were provided by one of the ISP groups. (*"As vendors we felt it might be necessary to bribe the judges with food,"* quipped the spokesperson of the generous doughnut caterers.)

Each ISP group was presented an award according to its place in the final rankings by the judges. Although such awards were just small tokens in recognition of tremendous effort, the knowledge, understanding, and experience of building and evaluating even a rudimentary search engine hopefully will remain with each student who participated in the project.

Chapter 9

Further Reading

Many sources went into *Understanding Search Engines.* Some of these sources deserve special mention not only because of their influence on our book, but because they also provide additional and more complete understanding of specific topics surrounding search engines.

9.1 General Textbooks on IR

In the Data and Information Management course which provided some of the impetus for writing this book, the textbook used was Gerald Kowalski's, *Information Retrieval Systems: Theory and Implementation* [33]. This general purpose book looks at many basic concepts of IR. It contains good examples of data structures, background material on indexing, and cataloging while covering important areas such as item normalization and clustering as it relates to generating a thesaurus.

Another all-purpose book *Information Storage and Retrieval* by Robert R. Korfhage [32] contains comprehensive chapters on query structures and document file preparation. In the *next iteration* of the Data and Information Management course taught in the spring of 1999, the Korfhage textbook was used. In contrast to the previously mentioned books, *Managing Gigabytes: Compressing and Indexing Documents and Images* by I. H. Witten, A. Moffat, and T. C. Bell [63] is understandably more focused on storage issues, but it does address the subject of IR (indexing, queries, and index construction), albeit from a unique compression perspective.

One of the first books that covers various IR topics was actually a collection of survey papers edited by William B. Frakes and Ricardo Baeza-Yates. Their 1992 book [23], *Information Retrieval: Data Structures & Algorithms,* contains several seminal works in this area including the use of *signature-based* text retrieval methods by Christos Faloutsos, and the development of ranking algorithms by Donna Harman.

9.2   Computational Methods and Software

Two *SIAM Review* articles (Berry, Dumais, and O'Brien in 1995 [7] and Berry, Drmač, and Jessup in 1999 [6]) demonstrate the use of linear algebra for vector space IR models such as LSI. The latter of these articles [6] would be especially helpful to undergraduate students in applied mathematics or scientific computing. Details of matrix decompositions such as the QR factorization and SVD are found in the popular reference book [26], *Matrix Computations*, by Gene Golub and Charles Van Loan. The work of Kolda and O'Leary [31] demonstrates the use of alternative decompositions such as the SDD for vector space IR models. Recent work by Simon and Zha [56] on updating LSI models demonstrates how to maintain accurate low-rank approximations to term-by-document matrices in the context of dynamic text collections. Berry and Fierro in [11] discuss updating LSI in the context of the ULV (or URV) matrix decomposition.

In order to compute the SVD of large sparse term-by-document matrices, iterative methods such as Arnoldi in [38], Lanczos in [34, 47], subspace iteration [48, 47], and trace minimization [52] can be used. In [9], Berry discusses how the last three methods are implemented within the software libraries SVDPACK (Fortran 77) [4] and SVDPACKC (ANSI C) [5], which are available in the public domain.

Simple descriptions with Matlab examples for Lanczos-based methods are available in [2], and a good survey of public-domain software for Lanczos-type methods is available in [10]. Whereas most of the iterative methods used for computing the SVD are serial in nature, an interesting asynchronous technique for computing several of the largest singular triplets of a sparse matrix on a network of workstations is described in [62].

The work of Lehoucq [37] and Lehoucq and Sorensen [38] on the Arnoldi method should be considered for computing the SVD of sparse matrices. Their software is available in ARPACK [37]. If one ignores sparsity altogether or must compute the (truncated) SVD of relatively dense matrices, the LAPACK [1] and ScaLAPACK [13] software libraries are available. A quick summary of available software packages mentioned above along with

their corresponding websites is provided in Table 9.1. The LSI website listed
in Table 9.1 allows the user to search a few sample text collections using
LSI. Some of the client–server designs used in recent LSI implementations
are discussed in [39].

Table 9.1: Useful websites for software and algorithm descriptions.

URL	Description
http://www.cs.utk.edu/~lsi	LSI Website (papers, sample text collections, and client–servers for demonstrations)
http://www.netlib.org/svdpack	SVDPACK (Fortran-77) and SVDPACKC (ANSI C) libraries [4, 5]
http://www.netlib.org/templates	Postscript and HTML forms of [2] along with Fortran, C++, and Matlab software
http://www.netlib.org/lapack	Latest Users' Guide [1] and Fortran-77, C, C++ software
http://www.netlib.org/scalapack	Latest ScaLAPACK User's Guide [13] and software and Parallel ARPACK (PARPACK) software
http://www.netlib.org/scalapack/arpack96.tgz	Implicitly Restarted Arnoldi software in Fortran-77 [37]

9.3   Search Engines

One of the toughest tasks in writing about this subject is staying current on
commercial search engines. What we wrote today is not necessarily going
to be true next week, so we recommend bookmarking Search Engine Watch
(see Table 9.2) edited by Danny Sullivan. Search Engine Watch tracks the
major search engines, their latest features, and how to better utilize them
both as a user and a webmaster. For a small fee subscribers can access
additional information and receive biweekly, electronic updates.

Table 9.2: Useful websites for current search engines and user interface designs.

URL	Description
http://www.searchenginewatch.com	Current information on commercial search engines
http://www.aw.com/DTUI	Companion to [54] with resources and links
http://www.dlib.org/dlib/january97/ retrieval/01shneiderman.html	Abbreviated overview of user interfaces

9.4   User Interfaces

Ben Shneiderman's book [54], *Designing the User Interface: Strategies of Effective Human-Computer Interaction*, is more than a summary of the state of HCI. The book, now in its third edition, has a kind of renaissance quality about it. Shneiderman is not content to just write specifically about only computer interfaces but takes a broad widespread approach covering topics ranging from interface models to specific design guidelines that makes the book more enjoyable and relevant. Related to this book are the WWW sites listed in Table 9.2.

A similar book is Gary Marchionini's *Information Seeking in Electronic Environments* [40]. Marchionini's approach identifies users' information seeking patterns and then tries to match them to current information retrieval technologies.

Bibliography

[1] E. ANDERSON, Z. BAI, C. BISCHOF, J. DEMMEL, J. DONGARRA, J. D. CRUZ, A. GREENBAUM, S. HAMMARLING, A. MCKENNEY, S. OSTROUCHOV, AND D. SORENSEN, *LAPACK Users' Guide*, second ed., SIAM, Philadelphia, PA, 1995.

[2] R. BARRETT, M. BERRY, T. CHAN, J. DEMMEL, J. DONATO, J. DONGARRA, V. EIJKHOUT, R. POZO, C. ROMINE, AND H. VAN DER VORST, *Templates for the Solution of Linear Systems: Building Blocks for Iterative Methods*, SIAM, Philadelphia, 1994.

[3] D. BERG, *A Guide to the Oxford English Dictionary*, Oxford University Press, Oxford, 1993.

[4] M. BERRY, *SVDPACK: A Fortran 77 Software Library for the Sparse Singular Value Decomposition*, Tech. Report CS–92–159, University of Tennessee, Knoxville, TN, June 1992.

[5] M. BERRY, T. DO, G. O'BRIEN, V. KRISHNA, AND S. VARADHAN, *SVDPACKC: Version 1.0 User's Guide*, Tech. Report CS–93–194, University of Tennessee, Knoxville, TN, October 1993.

[6] M. W. BERRY, Z. DRMAČ, AND E. R. JESSUP, *Matrices, vector spaces, and information retrieval*, SIAM Review, 41 (1999), pp. 335–362.

[7] M. BERRY, S. DUMAIS, AND G. O'BRIEN, *Using linear algebra for intelligent information retrieval*, SIAM Review, 37 (1995), pp. 573–595.

[8] M. BERRY, B. HENDRICKSON, AND P. RAGHAVAN, *Sparse matrix reordering schemes for browsing hypertext*, in Lectures in Applied Mathematics Vol. 32: The Mathematics of Numerical Analysis, J. Renegar,

M. Shub, and S. Smale, eds., American Mathematical Society, Providence, RI 1996, pp. 99–123.

[9] M. W. BERRY, *Large scale singular value computations*, International Journal of Supercomputer Applications, 6 (1992), pp. 13–49.

[10] M. W. BERRY, *Survey of public-domain Lanczos-based software*, in Proceedings of the Cornelius Lanczos Centenary Conference, J. Brown, M. Chu, D. Ellison, and R. Plemmons, eds., SIAM, Philadelphia, PA, 1997, pp. 332–334.

[11] M. W. BERRY AND R. D. FIERRO, *Low-rank orthogonal decompositions for information retrieval applications*, Numerical Linear Algebra with Applications, 3 (1996), pp. 301–328.

[12] K. BHARAT AND A. BRODER, *Estimating the Relative Size and Overlap of Public Web Search Engines*, in 7th International World Wide Web Conference, Paper FP37, Elsevier Science, New York, 1998.

[13] L. S. BLACKFORD, J. CHOI, A. CLEARY, E. D'AZEVEDO, J. DEMMEL, I. DHILLON, J. DONGARRA, S. HAMMARLING, G. HENRY, A. PETITET, K. STANLEY, D. WALKER, AND R. C. WHALEY, *ScaLAPACK Users' Guide*, first ed., SIAM, Philadelphia, PA, 1997.

[14] A. BOOKSTEIN, *On the perils of merging Boolean and weighted retrieval systems*, Journal of the American Society for Information Science, 29 (1978), pp. 156–158.

[15] A. BOOKSTEIN, *Probability and fuzzy set applications to information retrieval*, Annual Review of Information Science and Technology, 20 (1985), pp. 117–152.

[16] C. BUCKLEY, A. SINGHAL, M. MITRA, AND G. SALTON, *New retrieval approaches using SMART: TREC 4*, in Proceedings of the Fourth Text Retrieval Conference (TREC-4), D. Harman, ed., Gaithersburg, MD, 1996, Department of Commerce, National Institute of Standards and Technology. NIST Special Publication 500-236.

[17] C. CLEVERDON, *Optimizing convenient online access to bibliographic databases*, in Document Retrieval Systems, P. Willett, ed., Taylor Graham, London, 1988, pp. 32–41.

[18] S. DEERWESTER, S. DUMAIS, G. FURNAS, T. LANDAUER, AND R. HARSHMAN, *Indexing by latent semantic analysis*, Journal of the American Society for Information Science, 41 (1990), pp. 391–407.

[19] I. DUFF, R. GRIMES, AND J. LEWIS, *Sparse matrix test problems*, ACM Transactions on Mathematical Software, 15 (1989), pp. 1–14.

[20] S. T. DUMAIS, *Improving the retrieval of information from external sources*, Behavior Research Methods, Instruments, & Computers, 23 (1991), pp. 229–236.

[21] C. ECKART AND G. YOUNG, *The approximation of one matrix by another of lower rank*, Psychometrika, 1 (1936), pp. 211–218.

[22] C. FALOUTSOS, *Signature files*, in Information Retrieval: Data Structures & Algorithms, W. B. Frakes and R. Baeza-Yates, eds., Prentice–Hall, Englewood Cliffs, NJ, 1992, pp. 44–65.

[23] W. B. FRAKES AND R. BAEZA-YATES, *Information Retrieval: Data Structures & Algorithms*, Prentice–Hall, Englewood Cliffs, NJ, 1992.

[24] R. FUNG AND B. D. FAVERO, *Applying Bayesian networks to information retrieval*, Communications of the ACM, 58 (1995), pp. 27–30.

[25] F. GEY, *Inferring probability of relevance using the method of logistic regression*, in Proceedings of the Seventeenth Annual ACM-SIGIR Conference, W. B. Croft and C. van Rijsbergen, eds., London, 1994, Springer-Verlag, Berlin, New York, pp. 222–241.

[26] G. GOLUB AND C. VAN LOAN, *Matrix Computations*, third ed., Johns Hopkins University Press, Baltimore, 1996.

[27] D. HARMAN, *Ranking algorithms*, in Information Retrieval: Data Structures & Algorithms, W. B. Frakes and R. Baeza-Yates, eds., Prentice–Hall, Englewood Cliffs, NJ, 1992, pp. 363–392.

[28] D. HARMAN, ED., *Proceedings of the Third Text Retrieval Conference (TREC-3)*, Gaithersburg, MD, 1995, Department of Commerce, National Institute of Standards and Technology. NIST Special Publication 500-225.

[29] D. HARMAN, *Relevance feedback revisited*, in Proceedings of the Fifteenth Annual International ACM SIGIR Conference on Research and Development in Information Retrieval, Copenhagen, Denmark, June 21–24, 1992, pp. 1–10.

[30] W. JONES AND G. FURNAS, *Pictures of relevance: A geometric analysis of similarity measures*, Journal of the American Society for Information Science, 38 (1987), pp. 420–442.

[31] T. G. KOLDA AND D. P. O'LEARY, *A semi-discrete matrix decomposition for latent semantic indexing in information retrieval*, ACM Transactions on Information Systems, 16 (1998), pp. 322–346.

[32] R. R. KORFHAGE, *Information Storage and Retrieval*, John Wiley & Sons, Inc., New York, 1997.

[33] G. KOWALSKI, *Information Retrieval Systems: Theory and Implementation*, Kluwer Academic Publishers, Boston, MA, 1997.

[34] C. LANCZOS, *An iteration method for the solution of the eigenvalue problem of linear differential and integral operators*, Journal of Research of the National Bureau of Standards, 45 (1950), pp. 255–282.

[35] R. R. LARSON, *Evaluation of advanced retrieval techniques in an experimental online catalog*, Journal of the American Society for Information Science, 43 (1992), pp. 34–53.

[36] S. LAWRENCE AND C. L. GILES, *Searching the world wide web*, Science, 280 (1998), pp. 98–100.

[37] R. B. LEHOUCQ, *Analysis and Implementation of an Implicitly Restarted Arnoldi Iteration*, Ph.D. thesis, Rice University, Houston, TX, 1995.

[38] R. B. LEHOUCQ AND D. C. SORENSEN, *Deflation techniques for an implicitly restarted Arnoldi iteration*, SIAM Journal on Matrix Analysis and Applications, 17 (1996), pp. 789–821.

[39] T. A. LETSCHE AND M. W. BERRY, *Large-scale information retrieval with latent semantic indexing*, Information Sciences, 100 (1997), pp. 105–137.

[40] G. MARCHIONINI, *Information Seeking in Electronic Environments*, Cambridge University Press, New York, 1995.

[41] THE MATHWORKS INC., *Using Matlab*, Natick, MA, 1998. Version 5.

[42] S. MAZE, D. MOXLEY, AND D. SMITH, *Authoritative Guide to Web Search Engines*, Neal-Schuman Publishers, Inc., New York, 1997.

[43] L. MIRSKY, *Symmetric gauge functions and unitarily invariant norms*, The Quarterly Journal of Mathematics, 11 (1960), pp. 50–59.

[44] NATIONAL LIBRARY OF MEDICINE, *National Library Medicine Fact Sheet: UMLS Metathesaurus.* http://www.nlm.nih.gov/pubs /factsheets/umlsmeta.html, December 14, 1998.

[45] G. O'BRIEN, *Information Management Tools for Updating an SVD-Encoded Indexing Scheme*, Master's thesis, University of Tennessee, Knoxville, TN, 1994.

[46] NATIONAL LIBRARY OF MEDICINE, *Medical subject headings — Annotated alphabetic list* 1997. U.S. Department of Health and Human Services, Bethesda, MD, August 1997.

[47] B. PARLETT, *The Symmetric Eigenvalue Problem*, Prentice–Hall, Englewood Cliffs, NJ, 1980.

[48] H. RUTISHAUSER, *Simultaneous iteration method for symmetric matrices*, Numerische Mathematik, 16 (1970), pp. 205–223.

[49] G. SALTON AND C. BUCKLEY, *Term weighting approaches in automatic text retrieval*, Information Processing and Management, 24 (1988), pp. 513–523.

[50] G. SALTON AND C. BUCKLEY, *Improving retrieval performance by relevance feedback*, Journal of the American Society for Information Science, 41 (1990), pp. 288–297.

[51] G. SALTON AND M. MCGILL, *Introduction to Modern Information Retrieval*, McGraw–Hill, New York, 1983.

[52] A. H. SAMEH AND J. A. WISNIEWSKI, *A trace minimization algorithm for the generalized eigenvalue problem*, SIAM Journal on Numerical Analysis, 19 (1982), pp. 1243–1259.

[53] B. SHNEIDERMAN, *Designing the User Interface: Strategies of Effective Human-Computer Interaction*, second ed., Addison–Wesley, Reading, MA, 1992.

[54] B. SHNEIDERMAN, *Designing the User Interface: Strategies of Effective Human-Computer Interaction*, third ed., Addison–Wesley, Reading, MA, 1998.

[55] B. SHNEIDERMAN, D. BYRD, AND W. B. CROFT, *Clarifying Search: A User-Interface Framework for Text Searches*, D-Lib Magazine, (1997). Available on-line at http://www.dlib.org/dlib/january97/retrieval/01shneiderman.html.

[56] H. SIMON AND H. ZHA, *On Updating Problems in Latent Semantic Indexing*, Tech. Report CSE-97-011, The Pennsylvania State University, University Park, PA, 1997.

[57] A. SINGHAL, G. SALTON, M. MITRA, AND C. BUCKLEY, *Document Length Normalization*, Tech. Report TR95-1529, Department of Computer Science, Cornell University, Ithaca, NY, 1995.

[58] K. SPARCK JONES, *A statistical interpretation of term specificity and its applications in retrieval*, Journal of Documentation, 28 (1972), pp. 11–21.

[59] D. SULLIVAN, *How Yahoo Works*. http://www.searchenginewatch.com, August 1, 1998.

[60] R. TYNER, *Sink or Swim: Internet Search Tools & Techniques*. Version 3.0, http://www.ouc.bc.ca/libr/connect96/search.htm, July 24, 1998.

[61] C. VAN RIJSBERGEN, *Information Retrieval*, second ed., Butterworths, London, 1979.

[62] S. VARADHAN, M. W. BERRY, AND G. H. GOLUB, *Approximating dominant singular triplets of large sparse matrices via modified moments*, Numerical Algorithms, 13 (1996), pp. 123–152.

[63] I. H. WITTEN, A. MOFFAT, AND T. C. BELL, *Managing Gigabytes: Compressing and Indexing Documents and Images*, Van Nostrand Reinhold, New York, 1994.

[64] R. S. WURMAN, *Information Architects*, Graphis Press Corporation, Zurich, 1996.

Index

<COMMENT> tags, 14
<META> tags, 14

AltaVista, 19, 82
anthropomorphism, 82, 85
Arnoldi, 58, 102

Baeza-Yates, 101
banded, 40
basis, 47, 49, 62
Bayesian models, 5
Berry, 102
bitmap, 29
Boolean, 66, 67, 70
    operator, 9, 66, 67, 84
    query, 67
    search, 7, 67
Bureau of Labor Statistics, 87

C, 80, 92
C++, 80, 90, 92
Cleverdon, 3
clustering, 40, 44
column
    pivoting, 49–51
    space, 47, 53, 54, 61, 62
compressed column storage (CCS), 40
compressed row storage (CRS), 40
contiguous word phrase, 25, 26, 66, 69

controlled vocabularies, 38
coordinates, 44, 58
Cornell University, 22
cosine, 35, 39, 44, 51, 58
    threshold, 35, 52
course, xii, 85, 87, 101

data compression, 22
dense matrix, 60
diagonal, 50, 53
DIALOG, 83
dictionary, 23, 24, 32
Dippo, 87
disambiguation, 21
document file, 2, 13, 23
document purification, 14
Drmač, 102
Dumais, 102
dynamic collections, 61

Eckart and Young, 54
Elsevier Science, 17
EMBASE, 17
Euclidean distance, 5
Excite, 19, 82

Faloutsos, 102
Fierro, 102
folding-in, 61, 62
formal public identifier (FPI), 15, 16

113